Springer Series in Reliability Engineering

Series Editor

Hoang Pham, Industrial and Systems Engineering, Rutgers, The State University of New Jersey, Piscataway, NJ, USA

Today's modern systems have become increasingly complex to design and build, while the demand for reliability and cost effective development continues. Reliability is one of the most important attributes in all these systems, including aerospace applications, real-time control, medical applications, defense systems, human decision-making, and home-security products. Growing international competition has increased the need for all designers, managers, practitioners, scientists and engineers to ensure a level of reliability of their product before release at the lowest cost. The interest in reliability has been growing in recent years and this trend will continue during the next decade and beyond.

The Springer Series in Reliability Engineering publishes books, monographs and edited volumes in important areas of current theoretical research development in reliability and in areas that attempt to bridge the gap between theory and application in areas of interest to practitioners in industry, laboratories, business, and government.

Now with 100 volumes!

Indexed in Scopus and EI Compendex

Interested authors should contact the series editor, Hoang Pham, Department of Industrial and Systems Engineering, Rutgers University, Piscataway, NJ 08854, USA. Email: **hopham@soe.rutgers.edu, or Anthony Doyle, Executive Editor, Springer, London. Email: anthony.doyle@springer.com**.

Kuibin Zhou

Jet Fire Due to Gas Leakage

Dynamical Theory and Risk Assessment

Springer

Kuibin Zhou
College of Safety Science and Engineering
Nanjing Tech University
Nanjing, Jiangsu, China

ISSN 1614-7839   ISSN 2196-999X (electronic)
Springer Series in Reliability Engineering
ISBN 978-981-97-5328-4   ISBN 978-981-97-5329-1 (eBook)
https://doi.org/10.1007/978-981-97-5329-1

© The Editor(s) (if applicable) and The Author(s), under exclusive license to Springer Nature
Singapore Pte Ltd. 2024

This work is subject to copyright. All rights are solely and exclusively licensed by the Publisher, whether
the whole or part of the material is concerned, specifically the rights of translation, reprinting, reuse
of illustrations, recitation, broadcasting, reproduction on microfilms or in any other physical way, and
transmission or information storage and retrieval, electronic adaptation, computer software, or by similar
or dissimilar methodology now known or hereafter developed.
The use of general descriptive names, registered names, trademarks, service marks, etc. in this publication
does not imply, even in the absence of a specific statement, that such names are exempt from the relevant
protective laws and regulations and therefore free for general use.
The publisher, the authors and the editors are safe to assume that the advice and information in this book
are believed to be true and accurate at the date of publication. Neither the publisher nor the authors or
the editors give a warranty, expressed or implied, with respect to the material contained herein or for any
errors or omissions that may have been made. The publisher remains neutral with regard to jurisdictional
claims in published maps and institutional affiliations.

This Springer imprint is published by the registered company Springer Nature Singapore Pte Ltd.
The registered company address is: 152 Beach Road, #21-01/04 Gateway East, Singapore 189721,
Singapore

If disposing of this product, please recycle the paper.

Preface

Gas leakage and jet fire accidents pose significant threats to human safety, industrial facilities, communities, and the environment. Understanding the dynamics and consequences of these events is crucial for effective risk mitigation and emergency response planning, with the increasing use of high-pressure gas in various sectors. This book delves into the intricate details of high-pressure gas transient leakages and the combustion behavior of jet diffusion fires, providing a comprehensive exploration of these critical phenomena.

Through seven meticulously crafted chapters, this book offers a systematic examination of various aspects related to gas leakage and jet fire dynamics. Chapter 1 sets the stage by elucidating the widespread utilization of high-pressure gas and the imperative need to comprehend the characteristics of gas leakage and jet fire. Subsequent chapters delve into quantitative analyses, modeling approaches, combustion behavior discussions, and practical applications.

Chapter 2 serves as a foundation, delineating the dynamical processes of gas transient leakages with a rigorous application of thermodynamics principles. Key challenges in leakage modeling are identified and addressed, paving the way for the development of reliable models that can accurately predict gas behavior under different conditions. In Chap. 3, the focus shifts to the combustion behavior of jet diffusion fires, with a thorough examination of semi-empirical correlations and boundary conditions influencing flame geometry and radiative fraction. This chapter underscores the intricate interplay between leakage dynamics and combustion phenomena. Chapter 4 extends the discussion by presenting thermal radiation models and their application in predicting radiant heat flux from jet fires of different directions and scales. Through meticulous comparisons with experimental measurement, the efficacy of different thermal radiation models is evaluated, culminating in a comprehensive framework for predicting radiant heat flux in high-pressure jet fires.

Building upon the foundations in Chaps. 2–4, Chap. 5 proposes a theoretical framework for assessing the thermal damage to people and structures exposed to high-pressure jet fires. The framework incorporates the gas leakage model, jet flame behavior correlation, thermal radiation model, and thermal damage criteria. The uncertainties of the framework are quantified by sensitivity analysis and Monte Carlo

vi   Preface

simulation, and its robustness is validated through field test results. In Chap. 6, the
theoretical framework is applied to analyze real-world scenarios involving natural gas
transmission pipelines, highlighting its practical utility in assessing the consequences
of gas leakage incidents on different targets. Finally, Chap. 7 encapsulates the key
findings of this book and offers recommendations for future research endeavors in
the realm of gas leakage and jet fire dynamics.

It is my sincere hope that this book serves as a valuable resource for researchers,
engineers, safety professionals, and policymakers engaged in the design, operation,
and regulation of high-pressure gas systems. By enhancing our understanding of gas
leakage and jet fire dynamics, we can strive to foster safer environments and avert
catastrophic accidents.

I should acknowledge the National Natural Science Foundation of China, for
supporting my research project "Coupling Mechanism and Dynamical Behaviors of
Flow and Combustion of Jet Fire induced by High-pressure Leakage" in 2019–2022
(No. 51876088). In particular, I am indebted to my graduate students (Jiaoyan Liu,
Meng Liu, Yuzhu Wang, Yueqiong Wu, Xuan Nie, Mengya Zhou, Ruixing Dong,
etc.) who have contributed to this project. I welcome any comments or suggestions
on topics covered in this book.

Nanjing, China   Kuibin Zhou

Contents

1 Importance and Characteristics of Gas Leakage and Jet Fire 1
   1.1 High-Pressure Gas 2
   1.2 Characteristics of Gas Leakage 4
   1.3 Characteristics of Jet Fire 6
   1.4 The Organization of This Book 8
   References 9

2 Dynamical Process of Gas Leakage 11
   2.1 Introduction 11
   2.2 Gas Leakage Model 13
      2.2.1 Model Based on the Ideal Gas Equation of State 16
      2.2.2 Model Based on the Abel-Noble Equation of State 17
      2.2.3 Model Based on the Van Der Waals Equation of State 19
      2.2.4 The Optimal $v_{cr,0}$ for the Abel-Noble and Van Der
         Waals Gas Leakage Models 20
   2.3 Concept and Model of Notional Nozzle 21
   2.4 Physical Property of Common Leaked Gas 23
   2.5 Analysis on the Model Stability 25
   2.6 Model Validation 29
   2.7 Summary 33
   References 36

3 Flame Behavior of Jet Diffusion Fire 37
   3.1 Jet Diffusion Flame in Still Air 39
      3.1.1 Flame Length 39
      3.1.2 Lift-Off Distance 41
      3.1.3 Radiative Fraction 43
      3.1.4 Application of Flame Length, Lift-Off Distance
         and Radiative Fraction Correlations 47
   3.2 Jet Diffusion Flame in a Complex Boundary 53
      3.2.1 Effect of Cross Wind 53
      3.2.2 Effect of the Leakage Exit Shape 54

viii   Contents

        3.2.3   Effect of Obstacle or Impinging Jet Flame   56
        3.2.4   Effect of the Solid Particle   58
        3.2.5   Effect of the Pit   59
        3.2.6   Effect of the Underwater   61
    3.3 Summary   62
    References   62

4   **Radiant Heat Flux from Jet Diffusion Fire**   67
    4.1 Thermal Radiation Model of Jet Fires   68
        4.1.1   Point Source Radiation Model   68
        4.1.2   Multipoint Source Radiation Model   68
        4.1.3   Solid Flame Radiation Model   69
        4.1.4   Line Source Radiation Model   71
    4.2 Comparisons of Radiant Heat Flux Between Model
         Predictions   74
        4.2.1   Radiant Heat Flux of Small Vertical Jet Fires   74
        4.2.2   Radiant Heat Flux of Medium Horizontal Jet Fires   77
        4.2.3   Radiant Heat Flux of Medium Vertical Jet Fires   79
        4.2.4   Radiant Heat Flux of Small Inclined Jet Fires   81
    4.3 Application of Line Source Radiation Model   83
    4.4 Summary   84
    References   85

5   **Theoretical Framework for Calculating Jet Fire Risk Induced
    by High-Pressure Transient Leakage**   87
    5.1 Theoretical Framework of Gas Leakage, Jet Flame, Thermal
         Radiation and Damage Criteria   89
        5.1.1   Damage Criteria for People and Structures: Radiant
                Heat Flux Threshold   91
        5.1.2   Damage Criteria for People and Structures: Thermal
                Dose   93
        5.1.3   Damage Criteria for People and Structures: Target
                Temperature   94
    5.2 Parameter Sensitivity and Uncertainty Analysis of Input
         Parameters   95
        5.2.1   Parameter Sensitivity Analysis of the Theoretical
                Framework   96
        5.2.2   Uncertainty Analysis on the Results of the Theoretical
                Framework   97
    5.3 Validation of Theoretical Framework   100
        5.3.1   Case 1: Large Hydrogen Jet Fire   100
        5.3.2   Case 2: Large Natural Gas Jet Fire   103
        5.3.3   Case 3: Large Hydrogen/Natural Gas Mixture Jet Fire   107
    5.4 Summary   108
    References   110

Contents   ix

**6 Application of Theoretical Framework into the Jet Fire
   Consequence of Nature Gas Transmission Pipeline** 111
   6.1 Pipeline Incident in Tangshan, China 112
       6.1.1 Gas Leakage and Jet Fire 112
       6.1.2 Radiant Heat Flux and Injury/Damage 114
   6.2 Pipeline Incident in Ghislenghien, Belgium 116
       6.2.1 Gas Leakage and Jet Fire 117
       6.2.2 Radiant Heat Flux and Injury/Damage 117
   6.3 Summary 123
   References 124

7 Conclusions and Future Research 125
   7.1 Main Conclusions 125
   7.2 Recommendations for Future Research 127
   References 129

Chapter 1
Importance and Characteristics of Gas Leakage and Jet Fire

Contents

1.1	High-Pressure Gas	2
1.2	Characteristics of Gas Leakage	4
1.3	Characteristics of Jet Fire	6
1.4	The Organization of This Book	8
References		9

The pressure vessel and pipe have been widely used to store and convey gas fuel in the petroleum, petrochemical and chemical industries, as well as in the urban residential area. The imbalance between energy distribution and demand continues to drive the expansion of trans-regional and transnational pipelines for natural gas, as well as the development of underground caverns for natural gas storage. In addition, the climate change requires the increasing use of hydrogen that is usually stored at high pressure. The rupture of vessel and pipe causes the fast flow of leaked gas. Once the leaked gas is ignited by a potential hot source, a jet fire could form to cause a serious of disastrous events such as explosions. Moreover, the self-ignition is often reported due to the thermal or electrical energy produced by the gas leakage flow [1]. Take for example the fire accidents of natural gas pipelines whose statistical analysis shows the increase of self-ignition probability with the pipeline pressure and diameter [2]. In short, the leakage of combustible gas holds a large potential to induce a jet fire. The jet flame impingement and thermal radiation are the escalation vectors for the analysis of domino effect triggered by jet fires [3]. Figure 1.1 depicts the development of gas leakage and jet fire accidents.

Take for example the natural gas pipeline. Its potential leakage would cause a serious incident. In 2017, a pipeline in Alaska ruptured, leaking 210,000 to 310,000 cubic feet of gas a day at some points. The natural gas pipeline is underwater, and the floating ice prevented divers from reaching the site. In 2021, a blaze fed by a leaking gas pipeline in the Gulf of Mexico created the unimaginable image of an ocean on fire, nicknamed "the eye of fire". The disaster was only a short distance

---

© The Author(s), under exclusive license to Springer Nature Singapore Pte Ltd. 2024
K. Zhou, *Jet Fire Due to Gas Leakage*, Springer Series in Reliability Engineering,
https://doi.org/10.1007/978-981-97-5329-1_1

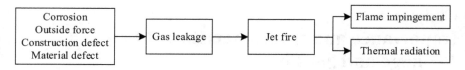

Fig. 1.1 Schematic for the accidental development of gas leakage and jet fire

from an oil platform. The natural gas pipeline rupture fire also ignites the vegetation and destroys the ecological function [4].

Another example is the underground gas storage facility whose leaks also pose a risk of a serious incident. In 2004, a failure led to the release of large amounts of natural gas, causing an explosion at an underground natural gas storage facility in Moss Bluff, Texas. In 2014, a leak also occurred at the Bergermeer gas storage facility in the Netherlands. Through October 23, 2015 to February 11, 2016, a significant methane leak incident occurred at the Aliso Canyon underground natural gas storage facility, with a release of approximately 97,100 tons of methane. Thousands of nearby residents were evacuated, and many reported health issues related to the natural gas leak, including headaches, stinging eyes, and vomiting. It should be stressed that these leaks had significant implications for global climate change, as methane is a greenhouse gas with a more potent warming effect than carbon dioxide.

In process industry, a historical survey on 84 jet fire accidents clarified that an additional event with much more severe effects and consequences was caused in 50% of the described cases [5]. In addition, the high-pressure gas leakage, fire and explosion accidents are also reported in manufacturing plants, transportation process, consumption sites and other scenarios. Figure 1.2 shows the numbers of accidents resulting in injuries or deaths and of deaths related to the High Pressure Gas Safety Act in Japan from 2003 to 2022. Accordingly, it is of great importance and interest to explore the gas leakage and jet fire.

1.1 High-Pressure Gas

High-pressure gas refers to a state in which a gas is subjected to pressures significantly above the ambient pressure. The exact threshold for defining high pressure can vary depending on the context and industry standards. In many cases, high pressure is above 1 MPa. High-pressure gases are utilized in various applications for specific reasons.

Compressing gases into a high-pressure state increases their density, which facilitates the efficient and economical storage or transportation of many gasses in a confined space or an industrial pipeline. Additionally, high-pressure conditions can alter reaction kinetics, which promotes some reactions and leads to increased productivity in chemical and industrial processes. Accordingly, High-pressure gases are commonly used in industrial processes such as manufacturing, chemical production,

1.1 High-Pressure Gas

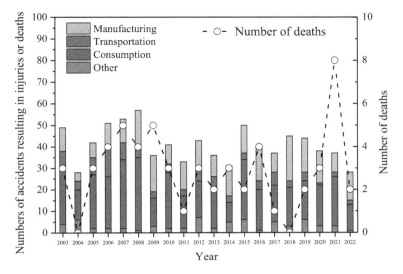

Fig. 1.2 The numbers of accidents resulting in injuries or deaths and of deaths related to the High Pressure Gas Safety Act in Japan (*Data source* the High Pressure Gas Safety Institute of Japan)

and metal fabrication, and they are encountered in oil and gas exploration, production, and refining processes. The storage and conveyance of high-pressure gases require the use of pressure vessels and pipes and underground caverns.

In pipeline transportation industry, the energy imbalance between distribution and demand still increases the multinational pipelines of natural gas, and the number of underground natural gas storage sites. Additionally, the climate change requires the increasing use of the hydrogen whose storage and transportation are often in a high-pressure state. The available natural gas pipelines can potentially support the transportation of the hydrogen in the future. Figure 1.3 shows the rapid development of the natural gas pipeline in Chinese mainland. In addition, the diameter and working pressure of the new natural gas pipeline also increase, as the pipeline steel and welding technology develops. Take for example the China-Russia east route natural gas pipeline project in which the pipeline diameter and pressure reach 1422 mm and 12 MPa, respectively.

In car industry, low emissions, high economic efficiency and safety are the fundamental requirement. The compressed natural gas (CNG) vehicles have been widely produced for the short-distance transportation and urban public transportation systems. Compared to gasoline vehicles, CNG vehicles do not emit sulfur compounds and lead in their exhaust, and significantly reduce the emissions of carbon monoxide, hydrocarbons, and nitrogen oxides. The hydrogen fuel cell vehicles whose storage pressure reaches 70 MPa, grows rapidly, for the achievement of zero emission. The vehicles powered by gaseous fuel, could cause jet fires from thermally activated pressure relief devices (TPRD) or storage vessel rupture before the action of TPRD, in the event of collision or fire. The thermal runaway of Lithium ion battery vehicles

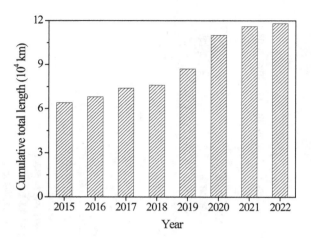

Fig. 1.3 The development of natural gas pipelines in Chinese mainland (*Data source* China Natural Gas Development Report)

also can generate high-pressure combustible gas and lead to gas leakage and jet fires. In addition, the wide use of CNG and hydrogen-powered vehicles also significantly increases the number of refueling stations.

With the increasing integration of renewable energy sources such as wind and solar power into the electricity grid, the need for effective energy storage solutions has become paramount. Compressed gas energy storage systems provide a viable option for storing excess energy generated during periods of low demand and releasing it during peak demand hours. The compressed gas is stored in underground caverns, depleted gas reservoirs, or aboveground tanks. In addition, certain medical procedures and equipment involve the use of high-pressure gases, and scientific research often utilizes high-pressure gases in experiments and testing. The expansion of high-pressure gases also may be involved in the propulsion system of space exploration and aerospace technology. In short, high-pressure gases are present in various natural and artificial scenarios.

1.2 Characteristics of Gas Leakage

Gas leaks can occur due to a variety of factors [6, 7]. One of the most common causes of gas leakage is the failure of materials and equipment used in gas containment and transportation systems. This category includes corrosion, wear and tear, manufacturing defects, and material incompatibility, etc. The aging and corrosion of pipelines were important factors contributing to the natural gas explosion accident in Shiyan in 2021. Secondly, improper installation and inadequate maintenance of gas systems significantly increase the risk of leaks, encompassing issues such as poor workmanship, lack of regular inspection, and delayed repairs. In the 2010 San Bruno

1.2 Characteristics of Gas Leakage   5

pipeline explosion, the pipeline ruptured due to a flawed weld and inadequate inspection and maintenance practices, as revealed by the National Transportation Safety Board investigation. Thirdly, human error also plays a crucial role in incidents of gas leakage, ranging from mistakes made during the installation and maintenance of gas systems to the incorrect operation of gas appliances. Examples include improper handling, operational mistakes, and a lack of awareness. In the 2004 Ghislenghien gas explosion, the rupture of a high-pressure gas pipeline resulted from the construction equipment striking the pipeline. On September 7, 2023, a high-pressure gas leakage accident occurred in the gasification workshop of Ordos Yiding Ecological Agriculture Development Co., Ltd., and the jet flow caused workers who were doing maintenance and inspection at a height to fall, resulting in 10 deaths and 3 injuries. Additionally, natural events can cause or contribute to gas leaks, including adverse environmental conditions, natural disasters, and soil movement. In the 2016 Enshi gas explosion of China, a landslide caused the West–East natural gas pipeline to rupture at 1 km from the exit of the Yuanjiawan Tunnel. In the 2017 Qinglong gas explosion of China, the continuous heavy rainfall caused the slope to collapse and slide down, rupturing the natural gas pipeline and triggering a leak. In short, the cause of gas leakage is complex, and its consequence is serious, which requires the deep insight into the gas leakage process.

There is a fundamental distinction between gas and liquid leakages. In the case of liquid leakage flow, physical properties such as density and temperature often remain constant. However, the assumption of constant physical properties is only applicable to gas or vapor leakage at low velocities (below 0.3 times the speed of sound in gas). From the perspective of energy conservation and conversion, the conservation of mechanical energy can explain the occurrence of liquid leakage, where the kinetic energy of liquid flow originates from either liquid pressure or liquid potential energy. Conversely, gas leakage requires quantitative description through the first law of thermodynamics, where the internal energy of the gas is transformed into the kinetic energy of gas flow.

During the conversion of internal energy into kinetic energy, friction also consumes internal energy. Therefore, gas leakages can be categorized into throttling and free expansion releases. When gas leaks through a small crack, significant frictional losses occur, and only a small portion of the internal energy is converted into kinetic energy, known as throttling release. In contrast, when most of the internal energy is converted into kinetic energy, the gas leakage is termed as free expansion release. Throttling release is slow, and heat exchange between the gas and surroundings can maintain a constant gas temperature, fitting an isothermal process. On the other hand, the assumption of adiabatic behavior often holds true for free expansion release of gas or vapor with rapid flow through a hole or crack. In fact, gas leakages typically exhibit behavior somewhere between isothermal and adiabatic flows. The equation of thermodynamic process and the equation of gas state can help to build the gas leakage model that predicts the flow velocity at the leakage exit [8–14].

The gas leakage leads to a jet flow outside the vessel or pipe. The flow characteristics mainly depends on the jet velocity and the leakage exit size and shape. The flow structure, velocity and concentration fields are the main focus of jet flow

6   1 Importance and Characteristics of Gas Leakage and Jet Fire

characteristics. The gas velocity and concentration decay along the jet axis, and their radial profiles transit from the top hat distribution to the Gaussian distribution along the jet axis [15]. In particular, the Mach number (Ma) at the leakage exit can divide the flow into the subsonic jet (Ma < 1), the sonic jet (Ma = 1) and the supersonic jet (Ma > 1). The flow is choked or under-expanded with normal shocks near the leakage exit. The normal shock is mainly characterized by the location and diameter of the Mach disc. The location can be quantified by the distance of the Mach disc to the leakage exit. The ratio of Mach disc diameter to Mach disc distance seems constant for a given gas, for large pressure ratios of vessel gas to ambient air [16, 17]. The ratio of Mach disc diameter to the leakage exit diameter is proportional to the square root of the pressure ratio [17]. More details on the choked jet flow can reference to the review work [18].

1.3 Characteristics of Jet Fire

In most of jet fire accidents, the combustion is diffusive. The mixing rate of fuel and air dominates the combustion rate of diffusive jet flames. The flow velocity at the leakage exit, as well as the flame buoyancy, largely influences the fuel–air mixing process [19]. As the jet velocity increases, the laminar flow transits to be the turbulent one. In a laminar jet flame, the mixing of fuel and air streams is relatively slow, for it is dominated by molecular diffusion. The laminar jet flame length is almost proportional to the jet flow. In the transition, the unsteady flutter appears near the jet flame tip, and develops into a noisy turbulent brush that moves towards the leakage exit. The transitional jet flame holds a little decrease in length as the jet velocity increases. In the fully turbulent jet flame, the mixing dominated by eddy diffusion is relatively rapid, and the flame length is independent on the jet velocity. Notice the analysis of Hottel and Hawthorne [19] only for jet flames attached the leakage exit.

A large enough jet velocity would quench the flame near the leakage exit, and the jet flame detaches the leakage exit, that is to say, a lift off phenomenon appears. In the lift-off region, the fuel jet stream mixes with the air, and a premixed fuel and air mixture forms to support the burning in the base of jet flame, which is validated by the appearance of blue flame in the lifted LPG jet fire [20]. If the premixed mixture is at the stoichiometric state, a premixed flame propagation model can help to explain the lift off phenomenon [21, 22]. In detail, the base of jet flame should stabilize at the interface where the local flow velocity of mixture gas equals its flame burning speed. If the premixed mixture is far from the stoichiometric state, the lift off phenomenon should be explained by the laminar diffusion flamelet extinction model that considers a turbulent jet flame to be a collection of laminar diffusion flamelet [23]. In detail, the flamelet quenches when its strain rate is larger than a critical scalar dissipation rate. The strain rate of flamelet can be quantified by the ratio of its characteristic flow velocity to length. Other theories such as the turbulent intensity theory, the large eddy concept and the edge-flame concept, are also proposed for the explanation of

1.3 Characteristics of Jet Fire

lift off phenomenon [24]. As the jet velocity continues to increase, the lift-off distance increases, and the mixture gas flow becomes unstable and oscillatory, which finally results in the blow off of flame.

The temperature distribution of jet fire also relies on the flame buoyancy and the flow velocity or momentum at the leakage exit. when the flow velocity is small, the effect of flame buoyancy is over that of flow momentum, and the temperature distribution along the flame centerline follows the correlation of McCaffrey [25]. In detail, the jet fire plume is divided into three zones. In the continuous flame zone, the centerline temperature keeps constant, while it decreases in the intermittent flame zone and more rapidly reduces in the plume zone, as shown in Fig. 1.4. However, as the jet velocity increases to enhance the effect of flow momentum, the centerline temperature distribution of jet fire would deviate from the McCaffrey's correlation. In particular, for the momentum-dominated jet fire, the flame temperature first increases and then decreases along the centerline, which can be well quantified by a second-degree polynomial [26]. In addition, the flame temperature first increases and then decreases as the heat release rate increases, and the critical heat release rate is 7 MW [26].

The flame buoyancy and the flow momentum also jointly affect the radiative fraction of jet flame. In general, the radiative fraction is constant for the buoyant jet fire, while it decreases for the momentum-dominated jet fire, as the jet velocity increases [27]. However, the total radiant heat output should increase as the jet velocity increases, for the total heat release rate always increases. The radiant heat, as well as the jet flame impingement, can cause serious damage on the nearby people, property and environment.

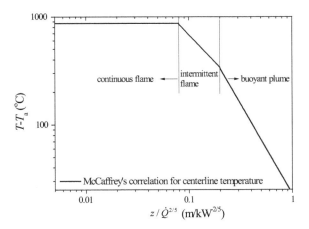

Fig. 1.4 McCaffrey's correlation for the centerline temperature of buoyant jet flames. T_a is the ambient air temperature, and $\dot{Q}$ is the heat release rate

1.4 The Organization of This Book

This book consists of seven chapters commencing with the first chapter introducing the wide use of high-pressure gas and the importance and characteristics of gas leakage and jet fire.

Chapter 2 quantitatively details the dynamical process of gas transient leakages using the first law of thermodynamics. Three key challenges are clarified for the leakage model development, including an equation of state to reliably quantify the behavior of high-pressure gas, an equation of thermodynamic process to accurately quantify the leakage process, and a model of notional nozzle to reasonably quantify the Mach disc in under-expanded leakages. Three high-pressure gas transient leakage models are detailed on the base of the ideal gas equation of state, the Abel-Noble equation of state and the van der Waals equation of state, respectively. Input parameters are given for leakage models. In particular, the critical pressure ratio should be optimized to distinguish the supersonic and subsonic leakages, for it affects the model stability. Finally, the model predictions are compared to the measured mass leakage rate, gas pressure, temperature and density of 90 MPa hydrogen transient leakages. The comparison clarifies the application conditions or limitations of models.

Chapter 3 intensively discusses the combustion behavior of jet diffusion fire. Different semi-empirical correlations are intensively reviewed for the flame length, lift-off distance and radiative fraction of turbulent free jet flames in still air. The leakage model outputs are input into the correlations to predict the flame geometry and radiative fraction against the experimental measurement. The optimal correlations are clarified to couple the transient leakage model. In addition, different boundary conditions including the cross wind, leakage exit shape, obstacle, solid particle, pit and underwater, are clarified to potentially affect the jet diffusion flame behavior, respectively.

Chapter 4 first introduces the point source radiation model, the multipoint source radiation model, the solid flame radiation model and the line source radiation model. Then the model predictions are compared to the measured radiant heat flux of vertical, horizontal and inclined jet fires of 1–10 m in length, by which the line source radiation model shows the best capacity. The line source radiation model, together with the leakage model, the correlations of flame length, lift-off distance and radiative fraction, is successfully to predict the measured radiant heat flux of 90 MPa hydrogen transient jet fires.

Chapter 5 proposes a theoretical framework to calculate the damage on people and structure due to the thermal radiation from high-pressure jet fire. The thermal damage criteria for people and structure is intensively reviewed. The input parameters by importance are ranked for the theoretical framework by parameter sensitivity analysis. The uncertainty of consequence calculated by the theoretical framework is stated by Monte Carlo simulation method. The robustness of the theoretical framework is justified by three field test results of the hydrogen, natural gas, and hydrogen/natural gas mixture jet fires. The maximum flame length reaches approximately 100 m in the test.

Chapter 6 uses the theoretical framework to quantify the jet fire consequence in two accidental cases of natural gas transmission pipelines. The pipeline condition parameters are input into the theoretical framework, for the calculation of leakage process, jet flame, radiant heat, and consequences of different targets. The damage consequences of targets are compared to those observed in the accidents.

In Chap. 7, the main conclusions are summarized for this book, and the recommendations are suggested for the future research on gas leakage and jet fire.

References

1. Kessler A, Schreiber A, Wassmer C et al (2014) Ignition of hydrogen jet fires from high pressure storage. Int J Hydrogen Energy 39(35):20554–20559
2. Acton MR, Baldwin PJ (2008) Ignition probability for high pressure gas transmission pipelines. In: Proceedings of the 7th international pipeline conference
3. Swuste P, van Nunen K, Reniers G et al (2019) Domino effects in chemical factories and clusters: an historical perspective and discussion. Process Saf Environ Prot 124:18–30
4. Scasta JD, Leverkus S, Tisseur D et al (2023) Vegetation response to a natural gas pipeline rupture fire in Canada's montane cordillera. Energy, Ecol Environ 8(5):457–470
5. Gómez-Mares M, Zárate L, Casal J (2008) Jet fires and the domino effect. Fire Saf J 43(8):583–588
6. Wang H, Duncan IJ (2014) Likelihood, causes, and consequences of focused leakage and rupture of U.S. natural gas transmission pipelines. J Loss Prev Process Ind 30:177–87
7. Vinnem JE, Røed W (2015) Root causes of hydrocarbon leaks on offshore petroleum installations. J Loss Prev Process Ind 36:54–62
8. Donaldson CD (1948) Note on the importance of imperfect-gas effects and variation of heat capacities on the isentropic flow of gases. NACA, Washington, DC, RM No. L8J14, pp 1–21
9. Li M, Zang L, Li C et al (2007) Analysis of pressure drop characteristics of deflation system under low atmospheric pressure environment. Acta Armamentarii 28(10):1234–1237
10. Li X, Bi J, Christopher DM (2013) Thermodynamic model of leaks from high-pressure hydrogen storage systems. J Tsinghua Univ (Sci & Tech) 53(04):503–508
11. Li X, Christopher DM, Bi J (2014) Release models for leaks from high-pressure hydrogen storage systems. Chin Sci Bull 59(19):2302–2308
12. Mohamed K, Paraschivoiu M (2005) Real gas simulation of hydrogen release from a high-pressure chamber. Int J Hydrogen Energy 30(8):903–912
13. Enkenhus KR (1967) On the pressure decay rate in the longshot reservoir. Von Karman Institute for Fluid Dynamics
14. Zhou K, Liu J, Wang Y et al (2018) Prediction of state property, flow parameter and jet flame size during transient releases from hydrogen storage systems. Int J Hydrogen Energy 43(27):12565–12573
15. Donaldson Cd, Snedeker RS (1971) A study of free jet impingement. Part 1. Mean properties of free and impinging jets. J Fluid Mech 45(2):281–319
16. Crist S, Glass DR, Sherman PM (1966) Study of the highly underexpanded sonic jet. AIAA J 4(1):68–71
17. Davidor W, Penner SS (1971) Shock standoff distances and Mach-disk diameters in underexpanded sonic jets. AIAA J 9(8):1651–1653
18. Franquet E, Perrier V, Gibout S et al (2015) Free underexpanded jets in a quiescent medium: a review. Prog Aerosp Sci 77:25–53
19. Hottel HC, Hawthorne WR (1948) Diffusion in laminar flame jets. Symp Combust & Flame & Explos Phenom 3(1):254–266

20. Kiran DY, Mishra DP (2007) Experimental studies of flame stability and emission characteristics of simple LPG jet diffusion flame. Fuel 86(10–11):1545–1551
21. Vanquickenborne L, van Tiggelen A (1966) The stabilization mechanism of lifted diffusion flames. Combust Flame 10(1):59–69
22. Kalghatgi GT (1984) Lift-off heights and visible lengths of vertical turbulent jet diffusion flames in still air. Combust Sci Technol 41(1–2):17–29
23. Peters N, Williams FA (1983) Liftoff characteristics of turbulent jet diffusion flames. AIAA J 21(3):423–429
24. Lyons KM (2007) Toward an understanding of the stabilization mechanisms of lifted turbulent jet flames: experiments. Prog Energy Combust Sci 33(2):211–231
25. McCaffrey BJ (1979) Purely buoyant diffusion flames: some experimental results. National Bureau of Standards, Washington, D.C.
26. Mercedes GM, Miguel M, Joaquim C (2009) Axial temperature distribution in vertical jet fires. J Hazard Mater 172(1):54
27. Zhou K, Liu J, Jiang J (2016) Prediction of radiant heat flux from horizontal propane jet fire. Appl Therm Eng 106:634–639

Chapter 2
Dynamical Process of Gas Leakage

Contents

2.1	Introduction	11
2.2	Gas Leakage Model	13
	2.2.1 Model Based on the Ideal Gas Equation of State	16
	2.2.2 Model Based on the Abel-Noble Equation of State	17
	2.2.3 Model Based on the Van Der Waals Equation of State	19
	2.2.4 The Optimal $v_{cr,0}$ for the Abel-Noble and Van Der Waals Gas Leakage Models	20
2.3	Concept and Model of Notional Nozzle	21
2.4	Physical Property of Common Leaked Gas	23
2.5	Analysis on the Model Stability	25
2.6	Model Validation	29
2.7	Summary	33
References	....	36

2.1 Introduction

The gas leakage is a typical thermodynamic process in which the internal energy of gas is transformed to the kinetic energy, which requires the equation of thermodynamic process and the first law of thermodynamics for the quantitative analysis. In addition, the quantitative description of gas behavior needs another important equation, i.e., the gas equation of state. Accordingly, the gas equation of state, the equation of thermodynamic process and the first law of thermodynamics are the theoretical base of the model development of the gas leakage. The gas leakage model functions as a bridge to link the gas parameters, including the thermal and mechanical parameters, inside the vessel (Level 1 in Fig. 2.1) and at the leakage exit (Level 2 in Fig. 2.1).

The leakage behavior can be divided into a steady release and an unsteady release, depending on the time variation of gas state properties inside the vessel. If the vessel

---

© The Author(s), under exclusive license to Springer Nature Singapore Pte Ltd. 2024
K. Zhou, *Jet Fire Due to Gas Leakage*, Springer Series in Reliability Engineering,
https://doi.org/10.1007/978-981-97-5329-1_2

Fig. 2.1 Schematic of choked flow near the leakage exit

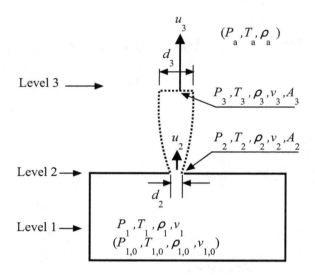

pressure and temperature always keep constant after a rupture, a steady leakage occurs. However, the accidental leakage is often unsteady due to the decrease of the vessel pressure with time. For the steady leakage, Chenoweth [1] extended the leakage model to the release of real gas, using the van der Waals equation of state to replace the ideal gas equation of state and Abel-Noble equation of state. For example, the air significantly deviates from behaving as an ideal gas to behaving as a real gas, when the stagnation pressures are of the magnitude of 50 atmospheres or greater [2]. The real gas accounts for the effect of molecular volume and intermolecular forces, which are not considered in the case of the ideal gas.

In the view of Woodward and Mudan [3], the time-varying leakage is inherently the decrease of discharge rate as the vessel pressure decays. The ideal gas equation of state is used to calculate the time-varying discharge rate, and thus an time-averaged discharge rate is proposed to represent the whole unsteady leakage [3]. Li et al. [4] further developed the transient leakage models for high-pressure gases using the ideal gas equation of state and the Abel-Noble equation of state, respectively. The comparison between the two models indicates that the model based on the Abel-Noble equation of state more accurately aligns with the numerical simulation. In recent, the Beattie-Bridgeman equation of state [5] and the van der Waals equation of state [6] were used to build the model of high-pressure gas transient leakage, respectively. In the above studies [3–6], the equation of an isentropic process is applied to quantitatively describe the leakage, assuming no heat loss or gain and a constant specific heat ratio in the thermodynamic process. In fact, the specific heat ratio varies with the gas temperature and pressure [7]. Accordingly, the assumption would fail, when there is a significant temperature difference between the gas tank and the surroundings.

2.2 Gas Leakage Model

In addition to the complex behavior of real gases, another challenge in high-pressure gas leakage is the transition from supersonic and choked flow to subsonic flow [4–6]. In the case of choked flow, the gas is under-expanded at the leakage exit, resulting in a pressure that exceeds the ambient pressure. The under-expanded gas jet expands immediately downstream of the exit in an attempt to equilibrate with the ambient conditions, which in turn increases the flow velocity. The continuous expansion of the gas gradually reduces its pressure until it equals the ambient pressure. Consequently, an interface with a Mach number of one, that is, a Mach disc, forms at Level 3, as depicted in Fig. 2.1. Accordingly, the concept and model of a notional nozzle are required to link the gas parameters at the leakage exit and those at the Mach disc [7, 8]. The conservation equations for mass and momentum provide the theoretical foundation for the notional nozzle model.

This chapter introduces the available models for gas transient leakage, as well as detailing the notional nozzle models for choked flow. It emphasizes the application conditions and limitations of the reviewed models. The focus of this chapter is not on the characteristics of gas jet flow, although there are numerous studies on the concentration, velocity and temperature fields of gas jet flows [9–12].

2.2 Gas Leakage Model

In the model development of gas leakage, the theoretical base includes the gas equation of state, the equation of thermodynamic process, the first law of thermodynamics and the equation of local sound speed. The ideal gas equation of state is

$$Pv = R_g T \tag{2.1}$$

in which P is the gas pressure, v is the gas specific volume and $v = 1/\rho$ in which ρ is the gas density, R_g is the gas constant and T is the gas absolute temperature.

The real gas equation of state should consider the molecular volume and the intermolecular forces. One widely-used equation for the real gas is the van der Waals equation expressed by

$$\left(P + a/v^2\right)(v - b) = R_g T \tag{2.2}$$

where a and b are the van der Waals constants, depending on the critical pressure and the critical temperature of gas. The term a/v^2 accounts for the intermolecular forces, while the co-volume b compensates for the finite volume occupied by the gas molecules. When the intermolecular attraction energy is little in comparison to the molecular kinetic energy, i.e., $a = 0$, Eq. (2.2) reverts to the so-called Able-Noble equation of state

$$P(v - b) = R_g T \tag{2.3}$$

14 2 Dynamical Process of Gas Leakage

After introducing the gas compressibility factor $Z = (1 - \rho b)^{-1}$, Eq. (2.3) can be rewritten as

$$Pv = ZR_gT \tag{2.4}$$

The gas compressibility factor depends on the reference pressure and temperature, which can be determined by generalized compressibility factor tables or diagrams. The reference pressure (P_r) and temperature (T_r) are defined as

$$P_r = P \big/ (P_c + 0.8), T_r = T \big/ (T_c + 0.8) \tag{2.5}$$

in which P_c and T_c are the critical pressure and temperature, respectively. It should be stressed that the co-volumes and their determination methods are different in Eqs. (2.2) and (2.3). In Eq. (2.3), the co-volume actually considers the coupled effect of the molecular volume and the intermolecular forces. The specific volume of real gas decreases as the intermolecular attractive force dominates, while it increases as the intermolecular repulsive force dominates. Accordingly, the co-volume in Eq. (2.3) could be larger or less than that in Eq. (2.2).

The equation of thermodynamic process can quantify the gas leakage process. It is of the form, respectively, for the ideal gas, the Able-Noble gas and the van der Waals gas

$$Pv^k = \text{constant} \tag{2.6}$$

$$P(v - b)^k = \text{constant} \tag{2.7}$$

$$(P + a/v^2)(v - b)^k = \text{constant} \tag{2.8}$$

where k is the polytropic exponent of thermodynamic process. In particular, the leakage process is isentropic and isothermal when k equals the specific heat ratio and one, respectively.

Application of the first law of thermodynamics into the leakage flow, the flow velocity at the leakage exit (u_2) can be calculated by

$$u_2 = \sqrt{2(h_1 - h_2)} = \sqrt{2c_p(T_1 - T_2)} \tag{2.9}$$

in which h is the gas enthalpy, and c_p is the gas specific heat at constant pressure. In Eq. (2.9), the flow velocity of gas is very little inside the vessel.

Substitution of Eqs. (2.6)–(2.8) into Eq. (2.9) can lead to the flow velocity, respectively, for the ideal gas, the Able-Noble gas and the van der Waals gas.

$$u_2 = \sqrt{\frac{2k}{k - 1} P_1 v_1 \left(1 - \left(\frac{P_2}{P_1}\right)^{\frac{k-1}{k}}\right)} \tag{2.10}$$

2.2 Gas Leakage Model
15

$$u_2 = \sqrt{\frac{2k}{k-1}P_1(v_1 - b)\left(1 - \left(\frac{P_2}{P_1}\right)^{\frac{k-1}{k}}\right)} \qquad (2.11)$$

$$u_2 = \sqrt{\frac{2k}{k-1}(P_1 + a/v_1^2)(v_1 - b)\left(1 - \left(\frac{P_2 + a/v_2^2}{P_1 + a/v_1^2}\right)^{\frac{k-1}{k}}\right)} \qquad (2.12)$$

The speed of sound in a gas (u_s) is defined as the speed at which a reversible pressure wave propagates. It is

$$u_s = \left(\frac{\partial P}{\partial \rho}\right)_s^{1/2} = v\left(-\frac{\partial P}{\partial v}\right)_s^{1/2} \qquad (2.13)$$

The partial derivative of both sides of Eqs. (2.6)–(2.8) with respect to specific volume, respectively, leads to

$$\left(\frac{\partial P}{\partial v}\right)_s = -k\frac{P}{v} \qquad (2.14)$$

$$\left(\frac{\partial P}{\partial v}\right)_s = -k\frac{P}{v-b} \qquad (2.15)$$

$$\left(\frac{\partial P}{\partial v}\right)_s = -k\frac{P + a/v^2}{v-b} + 2a/v^3 \qquad (2.16)$$

Finally, substitution of Eqs. (2.1)–(2.3) and (2.14)–(2.16) into Eq. (2.13) yields the speed of sound, respectively, in the ideal gas, the Able-Noble gas and the van der Waals gas.

$$u_s = \sqrt{kR_gT} \qquad (2.17)$$

$$u_s = \frac{v}{v-b}\sqrt{kR_gT} \qquad (2.18)$$

$$u_s = \sqrt{\frac{v^2}{(v-b)^2}kR_gT - \frac{2a}{v}} \qquad (2.19)$$

As indicated by the comparison between Eqs. (2.17)–(2.19), the co-volume b increases the speed of sound, while the intermolecular force decreases the sound speed.

2.2.1 Model Based on the Ideal Gas Equation of State

The choked leakage flow could occur when the vessel gas pressure is large enough. The appearance of choked flow indicates that the P_2/P_1 reaches a critical pressure ratio and that the flow velocity equals the sound speed at the leakage exit. Accordingly, Eqs. (2.6), (2.10) and (2.17) can derive the critical pressure ratio (v_{cr}), i.e.,

$$v_{cr} = \left(\frac{2}{k+1}\right)^{k/k-1} \tag{2.20}$$

If the ratio of the ambient pressure (P_a) to the vessel pressure (P_1) is larger than the critical pressure ratio, the leakage flow is subsonic. Accordingly, the pressure at the leakage exit (P_2) can be expressed by

$$P_2 = \begin{cases} v_{cr}P_1 & v_{cr}P_1 > P_a, \text{ chocked flow} \\ P_a & v_{cr}P_1 \le P_a, \text{ subsonic flow} \end{cases} \tag{2.21}$$

Substitution of Eqs. (2.20) and (2.21) into Eq. (2.10) leads to the formula of the gas flow velocity at the leakage exit

$$u_2 = \begin{cases} \sqrt{\frac{2k}{k+1}P_1 v_1} & v_{cr}P_1 > P_a \\ \sqrt{\frac{2k}{k-1}P_1 v_1 \left(1 - \left(\frac{P_a}{P_1}\right)^{\frac{k-1}{k}}\right)} & v_{cr}P_1 \le P_a \end{cases} \tag{2.22}$$

According to Eq. (2.6), the gas specific volume is, at the leakage exit

$$v_2 = \begin{cases} v_1\left(\frac{k+1}{2}\right)^{1/(k-1)} \text{ or } v_{1,0}\left(\frac{P_{1,0}}{v_{cr}P_1}\right)^{1/k} & v_{cr}P_1 > P_a \\ v_1\left(\frac{P_1}{P_a}\right)^{1/k} \text{ or } v_{1,0}\left(\frac{P_{1,0}}{P_a}\right)^{1/k} & v_{cr}P_1 \le P_a \end{cases} \tag{2.23}$$

Given the area of leakage exit (A_2), Eqs. (2.22) and (2.23) can give the mass flow rate ($\dot{q}_m$) at the leakage exit

$$\dot{q}_m = \frac{A_2 u_2}{v_2} = \begin{cases} A_2\sqrt{k\left(\frac{2}{k+1}\right)^{\frac{k+1}{k-1}}\frac{P_{1,0}}{v_{1,0}}\left(\frac{P_1}{P_{1,0}}\right)^{\frac{k+1}{k}}} & v_{cr}P_1 > P_a \\ A_2\left(\frac{P_a}{P_{1,0}}\right)^{\frac{1}{k}}\sqrt{\frac{2k}{k-1}\frac{P_{1,0}}{v_{1,0}}\left(\left(\frac{P_1}{P_{1,0}}\right)^{\frac{k-1}{k}} - \left(\frac{P_a}{P_{1,0}}\right)^{\frac{k-1}{k}}\right)} & v_{cr}P_1 \le P_a \end{cases} \tag{2.24}$$

Note the constant vessel volume during the whole leakage process. Therefore, the complete derivative of the vessel volume is zero, i.e., $d(m_1 v_1) = 0$. By the product rule, it leads to

2.2 Gas Leakage Model 17

$$\frac{dv_1}{v_1} = -\frac{dm_1}{m_1} = \frac{\dot{q}_m dt}{m_1} \tag{2.25}$$

The complete derivative of Eq. (2.6) leads to

$$\frac{dP_1}{P_1} = -k\frac{dv_1}{v_1} \tag{2.26}$$

The time variation of the gas specific volume, pressure, temperature and mass inside the vessel can be determined by the following iterative calculation

$$
\begin{aligned}
v_1(j+1) &= v_1(j)\left(1 + \frac{\dot{q}_m(j)}{m_1(j)}\Delta t\right) \\
P_1(j+1) &= P_1(j)\left(1 - k\frac{v_1(j+1) - v_1(j)}{v_1(j)}\right) \\
T_1(j+1) &= \frac{P_1(j+1) \cdot v_1(j+1)}{R_g} \\
m_1(j+1) &= m_1(j) - \dot{q}_m(j)\Delta t
\end{aligned}
\tag{2.27}
$$

Equations (2.20) and (2.24) can help to calculate the $\dot{q}_m(j+1)$ with the calculated $P_1(j+1)$ by Eq. (2.27). Equation (2.27) starts with $j = 1$, $v_1(1) = v_{1,0}$, $P_1(1) = P_{1,0}$ and $m_1(1) = m_{1,0}$, and does not end until P_1 decreases to equal P_a. Note the appearance of the transition from the choked to subsonic flows when $P_1 v_{cr}$ decreases to equal P_a. In Eq. (2.27), Δt is the time step of the iterative calculation. With the calculated v_1 and P_1 by Eq. (2.27), Eqs. (2.21)–(2.23) can help to compute the time variation of P_2, u_2, and v_2, during the whole leakage process.

2.2.2 Model Based on the Abel-Noble Equation of State

Equation (2.11) tells that the leakage flow velocity would increase to be the speed of sound, as the pressure ratio of P_2/P_1 increases to reach a critical value. Accordingly, the critical pressure ratio of Abel-Noble gas can be derived from Eqs. (2.7), (2.11) and (2.18). It is of the form

$$v_{cr} = \left[1 - \frac{(k-1)}{2}\frac{v_2^2(v_1 - b)^{k-1}}{(v_2 - b)^{k+1}}\right]^{\frac{k}{k-1}} \tag{2.28}$$

Note the decay of Eq. (2.28) to Eq. (2.20) when $b = 0$ accounts for no molecular volume. In addition, as indicated in Eqs. (2.20) and (2.28), the critical pressure is constant for the ideal gas, while it varies with the gas specific volume for the Able-Noble gas. Using the critical pressure ratio to distinguish the choked and subsonic flow periods, the gas pressure at the leakage exit can also expressed by Eq. (2.21).

18   2 Dynamical Process of Gas Leakage

With the substitution of Eqs. (2.21) and (2.28), Eq. (2.11) can be rewritten as

$$
u_2 = \begin{cases} \dfrac{v_2}{v_2-b}\sqrt{k\dfrac{P_1(v_1-b)^k}{(v_2-b)^{k-1}}} & P_1 v_{cr} > P_a \\[2ex] \sqrt{\dfrac{2k}{k-1}P_1(v_1-b)\left(1-\left(\dfrac{P_a}{P_1}\right)^{\frac{k-1}{k}}\right)} & P_1 v_{cr} \leq P_a \end{cases}
\tag{2.29}
$$

According to Eq. (2.7), the gas specific volume is, at the leakage exit

$$
v_2 = (v_1 - b)\left(\frac{P_1}{P_2}\right)^{1/k} + b
\tag{2.30}
$$

Equations (2.21), (2.29) and (2.30) can close the correlation of the mass flow rate at the leakage exit

$$
\dot{q}_m = \frac{A_2 u_2}{v_2}
\tag{2.31}
$$

The complete derivative of Eq. (2.7) leads to

$$
\frac{dP_1}{P_1} = -k\frac{dv_1}{v_1 - b}
\tag{2.32}
$$

With Eqs. (2.25) and (2.32), the time variation of the gas specific volume, pressure, temperature and mass inside the vessel can be determined by the following iterative calculation

$$
\begin{aligned}
v_1(j+1) &= v_1(j)\left(1 + \frac{\dot{q}_m(j)}{m_1(j)}\Delta t\right) \\
P_1(j+1) &= P_1(j)\left(1 - k\frac{v_1(j+1)-v_1(j)}{v_1(j)-b}\right) \\
T_1(j+1) &= \frac{P_1(j+1)\cdot\left[v_1(j+1)-b\right]}{R_g} \\
m_1(j+1) &= m_1(j) - \dot{q}_m(j)\Delta t
\end{aligned}
\tag{2.33}
$$

Equations (2.21) and (2.28)–(2.31) can help to calculate the $v_{cr}(j+1)$, $v_2(j+1)$, $P_2(j+1)$, $u_2(j+1)$ and $\dot{q}_m(j+1)$ with the calculated $v_1(j+1)$ and $P_1(j+1)$ by Eq. (2.33). Equation (2.33) starts with $j = 1$, $v_{cr}(1) = v_{cr,0}$, $v_1(1) = v_{1,0}$, $P_1(1) = P_{1,0}$ and $m_1(1) = m_{1,0}$, and does not end until P_1 decreases to equal P_a. Note the appearance of the transition from the choked to subsonic flows when $P_1 v_{cr}$ decreases to equal P_a. The critical pressure ratio could be calculated by Eq. (2.20) at the initial time, i.e., $v_{cr,0} = \left(2/(k+1)\right)^{k/k} - 1$.

2.2 Gas Leakage Model   19

2.2.3 Model Based on the Van Der Waals Equation of State

In comparison with the ideal gas and the Abel-Noble gas, the critical pressure ratio
of the van der Waals gas is defined as

$$\upsilon_{cr} = \frac{P_{cr} + a/v_{cr}^2}{P_1 + a/v_1^2} \tag{2.34}$$

The leakage flow is sonic when $P_2 + a/v_2^2 = P_{cr} + a/v_{cr}^2$. Equations (2.8), (2.12),
(2.19) and (2.34) can derive the critical pressure ratio

$$\upsilon_{cr} = \left[1 - \frac{(k-1)}{2} \frac{v_2^2(v_1-b)^{k-1}}{(v_2-b)^{k+1}} \right]^{\frac{k}{k-1}} \tag{2.35}$$

In the derivation of Eq. (2.35), the item $2a/v_2$ is little and neglected as compared
to $kR_g T_2 v_2^2 / (v_2 - b)^2$ in Eq. (2.19), for the gas pressure significantly reduces at the
leakage exit as compared to that inside the vessel. Equations (2.28) and (2.35) hold
the same form, but the definitions are different for the critical pressure ratios.

A similar method to Eq. (2.21) can also be suggested to distinguish the choked
and subsonic leakage flows of the van der Waals gas. It is of the form

$$P_2 + a/v_2^2 = \begin{cases} (P_1 + a/v_1^2)\upsilon_{cr} & (P_1 + a/v_1^2)\upsilon_{cr} > P_a + a'/v_a^2, \text{ chocked flow} \\ P_a + a'/v_a^2 & (P_1 + a/v_1^2)\upsilon_{cr} \le P_a + a'/v_a^2, \text{ subsonic flow} \end{cases} \tag{2.36}$$

Here, a and a' are the van der Waals constants for the leaked gas and the ambient air,
respectively. Substitution of Eqs. (2.34)–(2.36) into Eq. (2.12) leads to

$$u_2 = \begin{cases} \dfrac{v_2}{v_2-b} \sqrt{k \dfrac{(P_1+a/v_1^2)(v_1-b)k}{(v_2-b)^{k-1}}} & (P_1 + a/v_1^2)\upsilon_{cr} > P_a + a'/v_a^2 \\[4ex] \sqrt{\dfrac{2k}{k-1}(P_1 + a/v_1^2)(v_1 - b)\left(1 - \left(\dfrac{P_a+a'/v_a^2}{P_1+a/v_1^2}\right)^{\frac{k-1}{k}}\right)} & (P_1 + a/v_1^2)\upsilon_{cr} \le P_a + a'/v_a^2 \end{cases} \tag{2.37}$$

According to Eq. (2.8), the gas specific volume is, at the leakage exit

$$v_2 = (v_1 - b)\left(\frac{P_1 + a/v_1^2}{P_2 + a/v_2^2}\right)^{1/k} + b \tag{2.38}$$

Equations (2.35)–(2.38) can determine the mass flow rate at the leakage exit by
Eq. (2.31).

The complete derivative of Eq. (2.8) gives

$$\frac{d\left(P_1 + a/v_1^2\right)}{P_1 + a/v_1^2} = -k\frac{d(v_1)}{v_1 - b} \tag{2.39}$$

With Eqs. (2.25) and (2.39), the time variation of the gas specific volume, pressure, temperature and mass inside the vessel can be determined by the following iterative calculation

$$v_1(j+1) = v_1(j)\left(1 + \frac{\dot{q}_m(j)}{m_1(j)}\Delta t\right)$$

$$P_1(j+1) + a/v_1^2(j+1) = \left(P_1(j) + a/v_1^2(j)\right)\left(1 - k\frac{v_1(j+1) - v_1(j)}{v_1(j) - b}\right)$$

$$T_1(j+1) = \frac{\left[P_1(j+1) + a/v_1^2(j+1)\right] \cdot \left[v_1(j+1) - b\right]}{R_g} \tag{2.40}$$

$$m_1(j+1) = m_1(j) - \dot{q}_m(j)\Delta t$$

Equations (2.31) and (2.35)–(2.38) can help to calculate the $v_{cr}(j+1)$, $v_2(j+1)$, $P_2(j+1)$, $u_2(j+1)$ and $\dot{q}_m(j+1)$ with the calculated $v_1(j+1)$ and $P_1(j+1)$ by Eq. (2.40). Equation (2.40) starts with $j = 1$, $v_{cr}(1) = v_{cr,0}$, $v_1(1) = v_{1,0}$, $P_1(1) = P_{1,0}$ and $m_1(1) = m_{1,0}$, and does not end until $P_1 + a/v_1^2$ decreases to equal $P_a + a'/v_a^2$. Note the appearance of the transition from the choked to subsonic flows when $(P_1 + a/v_1^2)v_{cr}$ decreases to equal $P_a + a'/v_a^2$. The critical pressure ratio could be calculated by Eq. (2.20) at the initial time, i.e., $v_{cr,0} = \left(2/(k+1)\right)^{k/k} - 1$.

2.2.4 The Optimal $v_{cr,0}$ for the Abel-Noble and Van Der Waals Gas Leakage Models

In the leakage model of the Abel-Noble or van der Waals gas, the critical pressure ratio varies with the leakage time. The initially critical pressure ratio would significantly affect the stability of the model prediction in the incipient period. Accordingly, an optimal method is crucial to estimate the critical pressure ratio at the initial time. Here, the optimal method consists of an iterative computation, as described in detail in Fig. 2.2 and Fig. 2.3, respectively, for the Abel-Noble and van der Waals gases. The iterative calculation would converge, and the critical pressure ratio approaches to a constant. The constant is the optimal pressure ratio. If the iterative calculation diverges, the optimal method is done in vain. Section 2.5 will discuss the model stability in detail.

Fig. 2.2 Flow Chart for calculating the initially critical pressure ratio of the Abel-Noble gas

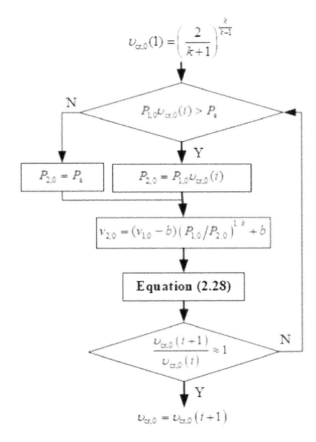

2.3 Concept and Model of Notional Nozzle

The notional nozzle, also called the notional source, is referred to as the pseudo diameter. The concept is proposed for the comparative analysis of the flow characteristics between the subsonic and choked jet flows. The flow characteristics relate to the flow structure, the concentration decay, and the velocity decay. The temperature decay could be another focus of the hot gas jet flow. With the concept of notional nozzle, the behavior of the choked jet is shown to be similar to the subsonic free jet, provided that the pseudo diameter is employed to describe the effective size of the jet source [7, 8]. In detail, the pseudo diameter and flow properties at the Mach disc, when substituted into the equations describing the subsonic free jet, would produce the concentration [7] and velocity [8] fields in the self-preserving region of the choked jet. Similarly, the correlations of subsonic jet flames have also been used to predict the supercritical jet flame behavior, by replacing the diameter and flow velocity at the leakage exit with those at the Mach disc [13]. It should be stressed

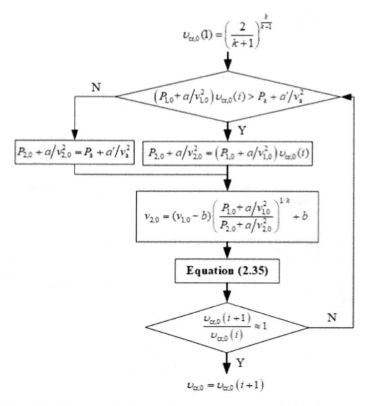

Fig. 2.3 Flow Chart for calculating the initially critical pressure ratio of the van der Waals gas

that the comparative analysis is meaningless in the development or expanding zone near the leakage exit.

The notional nozzle model helps to correlate the gas flow properties at the leakage exit and those at the Mach disc. Its aim is to calculate the pseudo diameter (d_3) and the flow velocity (u_3) at the Mach disc. In the development of the notional nozzle model, the following assumptions are generally made:

(1) There is no air entrainment to the gas jet through the notional nozzle boundary. That is to say, the mass flow rate equals between the leakage exit and Mach disc.
(2) The viscous force is negligible over the expansion surface, as compared to the pressure difference between the leakage exit and Mach disc.
(3) The pressure at the Mach disc equals the ambient pressure, while the temperatures at the Mach disc and at the leakage exit are approximately same.

Assumption (1) helps to build the conservation equation of mass between Levels 2 and 3. It is of the form

$$\rho_2 A_2 u_2 = \rho_3 A_3 u_3 \tag{2.41}$$

2.4 Physical Property of Common Leaked Gas

Assumption (2) aims to build the conservation equation of momentum to calculate the flow velocity at the Mach disc. For the ideal gas and the Abel-Noble gas, it can be expressed by.

$$\rho_2 A_2 u_2^2 - \rho_3 A_3 u_3^2 = A_2(P_3 - P_2) \tag{2.42}$$

For the van der Waals gas, it is of the form

$$\rho_2 A_2 u_2^2 - \rho_3 A_3 u_3^2 = A_2\left(P_3 + a/v_3^2 - P_2 - a/v_2^2\right) \tag{2.43}$$

Equations (2.41) and (2.42) or (2.43) can derive

$$u_3 = u_2 - \frac{(P_3 - P_2)}{\rho_2 u_2} \text{ or } u_3 = u_2 - \frac{P_3 + a/v_3^2 - P_2 - a/v_2^2}{\rho_2 u_2} \tag{2.44}$$

$$d_3 = d_2 \sqrt{\frac{u_2 \rho_2}{u_3 \rho_3}} \tag{2.45}$$

Assumption (3) tells $P_3 = P_a$ for the ideal gas and the Abel-Noble gas, and $P_3 + a/v_3^2 = P_a - a'/v_a^2$ for the van der Waals gas, and $T_3 = T_2$ for all gas. Accordingly, Eqs. (2.1)–(2.3) can be used to calculate the gas density at the Mach disc (ρ_3), for the ideal gas, the van der Waals gas and the Abel-Noble gas, respectively. In particular, Eqs. (2.44) and (2.45) reduces to be $u_3 = u_2$ and $d_3 = d_2$, as the choked flow decays and transitions to subsonic flow.

The surface friction results in a possible non-uniform velocity at the leakage exit, while the strong normal shock leads to a uniform velocity profile at the Mach disc. A discharge coefficient that can allow for the non-uniformity of the velocity profile, should be introduced for the term of $\rho_2 A_2 u_2^2$ in Eq. (2.41) to improve the model accuracy [7, 8]. In addition, the sound speed model was used to determine the flow velocities at both the leakage exit and the Mach disc [7]. In other words, Eqs. (2.17)–(2.19), instead of Eq. (2.43), are applied to calculate the flow velocity at Level 3, for the ideal gas, the Abel-Noble gas and the van der Waals gas, respectively. In essence, a different method to calculate the Mach disc indicates the difference in the definition of the notional nozzle.

2.4 Physical Property of Common Leaked Gas

The models in Sects. 2.2 and 2.3, can predict the state property and flow parameter, for the high-pressure leakage of pure or mixture gas, for they consider the physical property of gas. In detail, the molecular weight (M), critical pressure (P_c), critical temperature (T_c) and specific heat ratio (k), can make up the basic parameters that are input into the prediction model (Table 2.1).

The gas constant can be written as

24   2 Dynamical Process of Gas Leakage

Table 2.1 Physical parameters of common gas

Gas	M (kg/kmol)	P_c (MPa)	T_c (K)	k	R_g (J/(kg K))	a (m^6 Pa/kg^2)	b^a (m^3/kg)	b^b (m^3/kg)
H_2	2	1.3	33.2	1.4	4157	6084	0.0132	7.22×10^{-3}
CH_4	16	4.6	190.6	1.3	520	894	0.0027	6.49×10^{-4}
C_2H_6	30	4.88	305.4	1.3	277	619	0.0022	
C_3H_8	44	4.25	368.8	1.3	189	482	0.0020	3.79×10^{-4}
N_2	28	3.4	126.2	1.4	297	174	0.0014	
Air	29	3.77	132.5	1.4	287	468	0.0013	

^{a}It is the co-volume in Eq. (2.2)
^{b}It is the co-volume in Eq. (2.3)

$$R_g = R/M \tag{2.46}$$

in which R is the molar gas constant ($R = 8314$ J/(kmol K) and M is the molecular weight. The van der Waals constants are expressed by

$$a = \left(27R_g^2 T_c^2\right) \big/ (64P_c), \ b = \left(R_g T_c\right) \big/ (8P_c) \tag{2.47}$$

For the mixture gas consisting of η types of gas, the molecular weight can be calculated by

$$M = \sum_{i=1}^{\eta} y_i M_i \tag{2.48}$$

in which y_i is the volume fraction and the subscript i is the ith type of gas. And the critical pressure and temperature of the mixture gas can be calculated by

$$P_c = \sum_{i=1}^{\eta} y_i P_{c,i}, \ T_c = \sum_{i=1}^{\eta} y_i T_{c,i} \tag{2.49}$$

Thus, the gas constant and van der Waals constants can be determined by Eqs. (2.46)–(2.49) for the mixture gas. The specific heat ratio is the ratio of specific heat at constant pressure to that at constant volume, derived by

$$k = \sum_{i=1}^{\eta} w_i c_{p,i} \bigg/ \sum_{i=1}^{\eta} w_i c_{v,i} \tag{2.50}$$

where w_i is the mass fraction calculated by $w_i = y_i M_i \left/ \sum_{i=1}^{\eta} y_i M_i \right.$, and $c_{v,i}$ and $c_{p,i}$ are the specific heat at constant volume and pressure expressed by $c_{v,i} = R_{g,i} / (k_i - 1)$ and $c_{p,i} = k_i c_{v,i}$, respectively.

2.5 Analysis on the Model Stability

In theory, the leakage model can predict the variation of the thermodynamical and mechanical parameters with the leakage time, despite the initial gas pressure, temperature, volume and the gas type. However, the initial condition could affect the stability of model prediction. One important characteristic is the intensive fluctuation of prediction with the time in the incipient period, for the Abel-Noble and van der Waals gas leakage models. Here, the rated parameters of the on-board storage cylinder in the hydrogen and CNG (compressed natural gas) vehicle, as well as the storage tank in the hydrogen and CNG fueling station, are input into and test the leakage model of the Abel-Noble and van der Waals gas (Table 2.2).

Figure 2.4 shows the prediction of the thermodynamical and mechanical parameters as the hydrogen transiently releases from a cylinder of 35 MPa in pressure and 0.12 m^3 in volume. All the parameter curves are smooth predicted by the ideal gas leakage model. In comparison, except the parameters inside the cylinder, there is an intensive fluctuation for all the predicted parameters at the leakage exit and the Mach disc, for the Abel-Noble and van der Waals gas leakage models. As shown, the fluctuation period is short in the beginning and finally disappears with the birth of smooth curves. The fluctuation characteristics would pass to the parameters at the Mach disc through the notional nozzle model. Accordingly, it is important to explore the effect factor of the fluctuation behavior at the leakage exit.

As indicated in Fig. 2.5a–c, both the intensity and duration of the fluctuation increase as the initial hydrogen pressure increases from 35 to 90 MPa, especially for the van der Waals gas leakage model. An increase of the cylinder volume can also increase the fluctuation intensity and duration, as indicated by the comparison between Fig. 2.5c,d. In particular, the van der Waals gas leakage model is more sensitive to the initial pressure and the cylinder volume than the Able-Noble gas leakage model.

Table 2.2 Typical parameters of storage cylinder/tank in gas vehicle/fueling station

Location	Gas type	Volume (m^3)	Pressure (MPa)	Temperature (K)
On-board cylinder	Hydrogen	0.12	35/70	315
	CNG	0.05–0.15	20	315
Fueling station	Hydrogen	5	45/90	315
	CNG	6	40	315

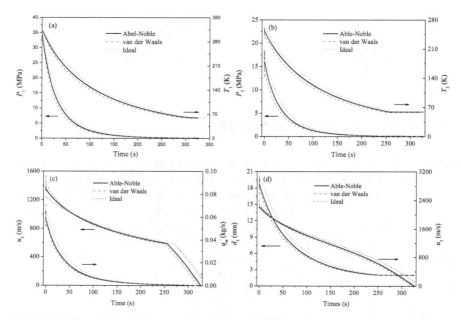

Fig. 2.4 Prediction of hydrogen transient leakage: **a** gas pressure and temperature inside the vessel, **b** gas pressure and temperature at the leakage exit, **c** gas flow velocity and mass flow rate at the leakage exit, and **d** Mach disc diameter and gas flow velocity at the Mach disc. The initial volume, pressure and temperature of hydrogen are $0.12 \, m^3$, 35 MPa and 315 K, respectively, and $d_2 = 2$ mm, and $v_{cr, 0} = 0.528$ by Eq. (2.20)

Figure 2.6 shows the comparison of the pressure and flow velocity fluctuation between two different exit diameters. In the van der Waals gas leakage model, an increase of the exit area can significantly reduce the time period of the parameter fluctuation, but cannot decrease the fluctuation intensity. Figure 2.7 shows the comparison of the pressure and flow velocity fluctuation between hydrogen and methane for the van der Waals gas leakage model. Comparatively speaking, the intensity and duration of fluctuation increase for the methane transient leakage. The stability of van der Waals model seems to enhance, as the molecular structure or molecular weight decreases.

As concluded from the above discussion, the ideal gas leakage model is always stable, while the other two model should consider the stability. Moreover, the leakage model of the van der Waals gas shows a more intensive fluctuation than that of the Able-Noble gas. The initial gas pressure, vessel volume and the molecular weight play a negative role in the stability of the van der Waals gas leakage model, while the leakage exit area would enhance the model stability. However, the fundamental factor to cause the fluctuation of parameters at the leakage exit, could be the critical pressure ratio at the initial time given by Eq. (2.20). The model stability issue could be solved by the optimal method in Sect. 2.2.4. As shown in Fig. 2.8, the time fluctuation of parameters at the leakage exit completely disappears when using the critical pressure

2.5 Analysis on the Model Stability

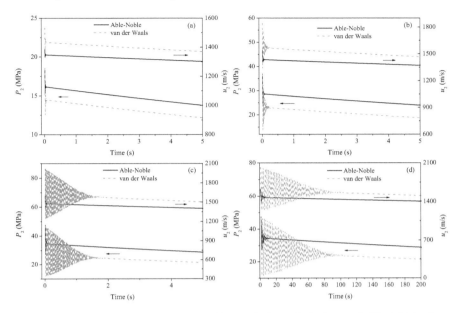

Fig. 2.5 Prediction of hydrogen pressure and flow velocity at the leakage exit. The initial volume and pressure are **a** 0.12 m^3 and 35 MPa, **b** 0.12 m^3 and 70 MPa, **c** 0.12 m^3 and 90 MPa, and **d** 5 m^3 and 90 MPa, respectively. The initial temperature is 315 K, and $d_2 = 2$ mm, and $\upsilon_{cr,0} = 0.528$ by Eq. (2.20)

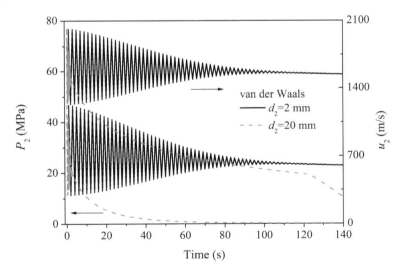

Fig. 2.6 Hydrogen pressure and flow velocity at the leakage exit predicted by the van der Waals gas leakage model. The initial volume and pressure are 5 m^3 and 90 MPa, respectively, and $\upsilon_{cr,0} = 0.528$ by Eq. (2.20)

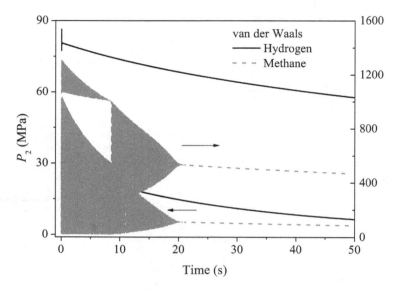

Fig. 2.7 Gas pressure and flow velocity at the leakage exit predicted by the van der Waals gas leakage model. The initial volume and pressure are 0.12 m^3 and 70 MPa, respectively, and $d_2 = $ 2 mm. $\upsilon_{cr,\,0} = 0.528$ and 0.546 by Eq. (2.20), respectively, for hydrogen and methane

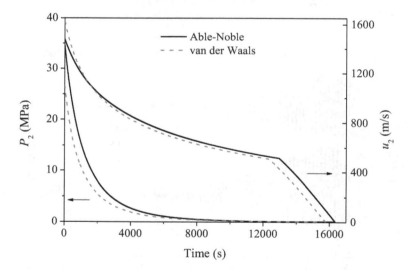

Fig. 2.8 Prediction of hydrogen pressure and flow velocity at the leakage exit. The initial volume, pressure and temperature are 5 m^3, 90 MPa and 315 K, respectively. $d_2 = 2$ mm, and $\upsilon_{cr,\,0} = 0.386$ and 0.303 determined by the optimal method, respectively, for the Abel-Noble and van der Waals gas

2.6 Model Validation 29

ratio calculated by the optimal method, even though the initial pressure reaches 90 MPa and the vessel volume is 5 m^3.

2.6 Model Validation

A series of experiments were recently conducted on the horizontal jet fires due to the transient release of 90 MPa hydrogen from a 25 L storage vessel [14]. In test, the leakage exit diameter varies from 1 to 3 mm. A pressure sensor (FGP, 0–1000 bar, 0–1000 Hz) was used to measure the gas pressure on the head of the vessel, and the temperature at the orifice exit was measured with K-type thermocouples in the upstream of the orifice. The mass flow rate of the gas was deduced by installing the tank on a numerical weighing device.

In order to validate the accuracy of the high-pressure gas leakage model, the measured gas pressure in vessel, temperature and mass flow rate at the orifice exit are used for comparison with the prediction. The initial pressure, temperature and volume of gas in the tank, as well as the leakage exit diameter, are the same for the three leakage models, while the initial gas specific volume and total mass are different and calculated based on the ideal gas, Abel-Noble, and van der Waals equation of state, respectively. Table 2.3 lists the initial parameters input into the model.

In general, the validation of theoretical model needs the accurate measurement data. The measurement accuracy significantly depends on the match of the sensor transient response with the characteristic variation of the leakage flow. The pressure sensor and K-type thermocouple are typical contact measurement methods, and could fail to fellow the transient variation of flow. In addition, the thermocouple would overestimate (or underestimate) the true fluid temperature due to the radiation gain from (or loss to) the hot (or cold) ambient surroundings. Accordingly, the following discussion is first on the mass flow rate at the leakage exit and the gas specific volume inside the vessel, and then on the gas pressure and temperature.

Figure 2.9 shows the variation of the mass flow rate with time at the leakage exit of 2 mm in diameter, for the comparison between the model prediction and experimental measurement. The ideal gas leakage model remarkably overestimates the hydrogen mass flow rate. As the gas molecular volume is considered, the prediction of Abel-Noble gas leakage model gradually approaches the measurement. The van der Waals gas leakage model agrees best with the measurement.

Table 2.3 The input parameters for all three models

Model	$P_{1,0}$ (MPa)	$T_{1,0}$ (K)	V (m^3)	d_2 (mm)	$v_{1,0}$ (m^3/kg)	$m_{1,0}$ (kg)
Ideal gas	90	315	0.025	1, 2, 3	0.0144	1.74
Abel-Noble	90	315	0.025	1, 2, 3	0.0220	1.14
van der Waals	90	315	0.025	1, 2, 3	0.0275	0.91

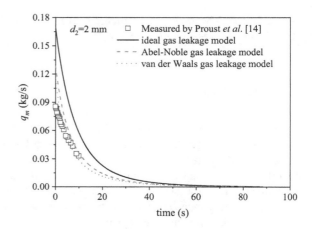

Fig. 2.9 Comparison of the mass leakage rate versus time between model prediction and experimental measurement

Figure 2.10 shows the hydrogen density versus pressure inside the vessel during the transient leakage, for the comparison between the predictions of different models against the measurement. The ideal gas leakage model always overestimates the gas density, despite the leakage exit diameter. In comparison, the van der Waals gas leakage model, as well as the Abel-Noble gas leakage model, can give a better prediction on the gas density.

Figure 2.11 shows the gas pressure versus leakage time inside the vessel predicted by the three different models. The ideal gas leakage model seems to well agree with the measurement. However, the other two models might underestimate the gas pressure inside the vessel to some extent, especially for the van der Waals gas leakage model. The reason could be the existence of a dynamic error when the pressure sensor is used for transient measurements [15]. Generally speaking, the simple relationship between the output signal ($y(t)$) and the input signal ($x(t)$) of pressure sensor can be expressed by correlation $y(t) = \beta_0 x(t - \tau_0)$ in which β_0 and τ_0 stand for the sensor sensitivity and response time, respectively. Obviously, the pressure sensor readings indicated by $y(t)$ is different from the true gas pressure indicated by $x(t)$.

Figure 2.12 presents the gas temperature versus leakage time inside the vessel predicted by the three different models. All the models significantly underestimate the experimental measurement. As compared to the measurement, the predicted gas temperature inside the vessel more rapidly decreases as the leakage time goes. The big underestimation could be explained by the assumption of the adiabatic leakage process. That is to say, there could exist a considerable heat transfer from the surroundings to the gas inside the vessel through the vessel wall, during the leakage process.

The variations of the gas temperature and mass flow rate with the pressure inside the vessel are used for a further analysis on the model accuracy. As shown in Fig. 2.13a, there is no difference the gas temperature versus the pressure inside the vessel between all model predictions. However, for the mass flow rate versus

2.6 Model Validation

Fig. 2.10 Comparison of the gas density versus pressure inside the vessel between model prediction and experimental measurement

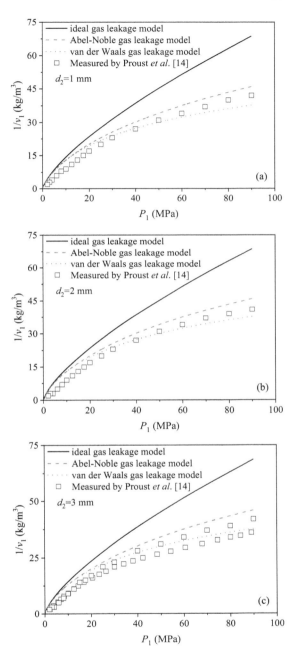

Fig. 2.11 Comparison of the gas pressure versus time between model prediction and experimental measurement inside the vessel

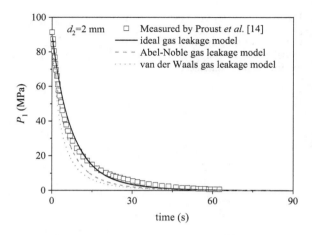

Fig. 2.12 Comparison of the gas temperature versus time between model prediction and experimental measurement inside the vessel

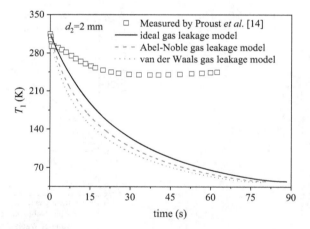

the vessel pressure, the prediction of van der Waals gas leakage model relatively approaches the experimental measurement, as indicated in Fig. 2.13b.

In Figs. 2.9, 2.10, 2.11, 2.12, and 2.13, all the model predictions are conducted with the assumption of the adiabatic leakage process. In the experimental test [14], the leakage process should be within the isentropic ($k = 1.4$, i.e. the specific heat ratio of hydrogen) and isothermal ($k = 1$) processes, for the leakage time is long enough for a considerable heat transfer to the hydrogen inside the vessel. Accordingly, instead of $k = 1.4$, the choice of $k = 1.1$ is made for the model predictions. As shown in Fig. 2.14, the ideal gas leakage model always overestimates the mass flow rate, hydrogen pressure, temperature and density inside the vessel, when the leakage time is less than 20 s. The vessel pressure is still high for the hydrogen to deviate far from ideal gas before the leakage time of 20 s. In comparison, the other two models, especially the van der Waals gas leakage model, can give a remarkable prediction.

2.7 Summary

Fig. 2.13 Comparison of the parameters inside the vessel between model prediction and experimental measurement: **a** gas temperature versus pressure and **b** mass flow rate versus gas pressure

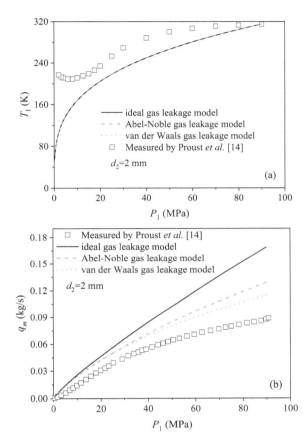

As can been seen from Fig. 2.14c, the leakage process should be simulated by the combination of $k = 1.1$ and 1.0 before and after approximately 20 s, respectively. In short, the high-pressure gas leakage is a complex thermodynamic process with a time-varying k.

2.7 Summary

The high-pressure gas leakage is a complexly physical process. In the quantitative modelling of gas leakage, the equation of state for the gas is used to quantify the behavior of the gas, while the equation of thermodynamic process and the first law of thermodynamics help to correlate the gas parameters inside the vessel and at the leakage exit. If the vessel pressure is high enough to form a choked leakage flow at

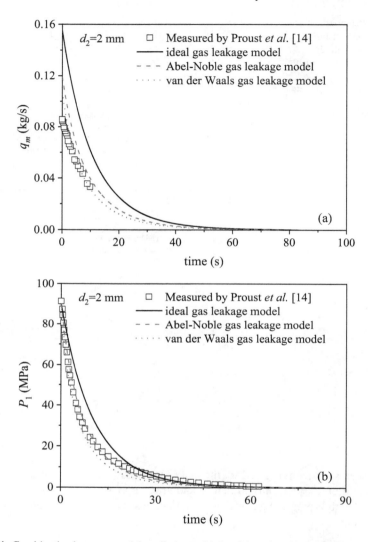

Fig. 2.14 Combination between model predictions with $k = 1.1$ against the experimental measurement: **a** mass flow rate versus time, **b** hydrogen pressure inside the vessel versus time, **c** hydrogen temperature inside the vessel versus time, and **d** hydrogen density versus pressure inside the vessel

the leakage exit, the equation of local sonic velocity is needed to distinguish the transition time from choked to subsonic flows.

As the gas pressure increases, the behavior of the gas will gradually deviate from that of an ideal gas. In detail, the volume of gas moleculars and the intermolecular forces play considerable roles in the gas volume and pressure, respectively. In addition to the ideal gas equation of state, the Abel-Noble and van der Waals equations of state are also recommended for developing the gas leakage model. The Abel-Noble

2.7 Summary

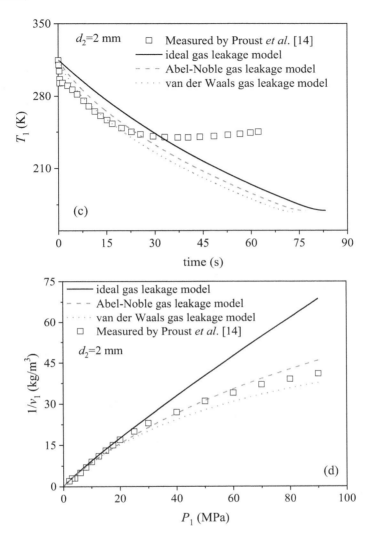

Fig. 2.14 (continued)

equation of state accounts only for the volume of gas molecules, while the van der Waals equation simultaneously considers both the volume of gas molecules and the intermolecular forces. Accordingly, three different leakage models are developed for the ideal gas, Abel-Noble gas and van der Waals gas, respectively. The three leakage models can predict the high-pressure leakage behavior of different gases and gas mixtures. In particular, one optimal method is proposed to determine the critical pressure ratio at the initial time for the Abel-Noble and van der Waals gas leakage models. The optimal pressure ratio aids in enhancing the model stability.

Data from a 90 MPa hydrogen transient leakage, including the mass leakage rate, gas pressure, temperature and density, are used to validate the model. A comparison between model predictions and experimental measurements shows that the van der Waals gas leakage model has greater robustness than the other two models. In fact, high-pressure gas transient leakage is significantly complex and poses challenges for accurate quantification, particularly due to a time-varying polytropic exponent in the thermodynamic process. Nonetheless, the outputs of the leakage model will be fed into the jet flame model described in Chap. 3.

References

1. Chenoweth DR (1983) Real gas results via the van der Waals equation of state and virial expansion extensions of its limiting Abel-Noble form. Sandia National Laboratories
2. Donaldson CD (1948) Note on the importance of imperfect-gas effects and variation of heat capacities on the isentropic flow of gases. NACA, Washington, DC, RM No. L8J14, pp 1–21
3. Woodward JL, Mudan KS (1991) Liquid and gas discharge rates through holes in process vessels. J Loss Prev Process Ind 4(3):161–165
4. Li X, Bi J, Christopher DM (2013) Thermodynamic model of leaks from high-pressure hydrogen storage systems. J Tsinghua Univ (Sci & Tech) 53(04):503–508
5. Mohamed K, Paraschivoiu M (2005) Real gas simulation of hydrogen release from a high-pressure chamber. Int J Hydrogen Energy 30(8):903–912
6. Zhou K, Liu J, Wang Y et al (2018) Prediction of state property, flow parameter and jet flame size during transient releases from hydrogen storage systems. Int J Hydrogen Energy 43(27):12565–12573
7. Birch AD, Brown DR, Dodson MG et al (1984) The structure and concentration decay of high pressure jets of natural gas. Combust Sci Technol 36(5–6):249–261
8. Birch AD, Hughes DJ, Swaffield F (1987) Velocity decay of high pressure jets. Combust Sci Technol 52(1–3):161–171
9. Becker HA, Hottel HC, Williams GC (1967) The nozzle-fluid concentration field of the round, turbulent, free jet. J Fluid Mech 30(2):285–303
10. Antonia RA, Prabhu A, Stephenson SE (1975) Conditionally sampled measurements in a heated turbulent jet. J Fluid Mech 72(3):455–480
11. Venkataramani KS, Tutu NK, Chevray R (1975) Probability distributions in a round heated jet. Phys Fluids 18(11):1413–1420
12. Birch AD, Brown DR, Dodson MG et al (1978) Turbulent concentration field of a methane jet. J Fluid Mech 88(3):431–449
13. Schefer RW, Houf WG, Williams TC et al (2007) Characterization of high-pressure, underexpanded hydrogen-jet flames. Int J Hydrogen Energy 32(12):2081–2093
14. Proust C, Jamois D, Studer E (2011) High pressure hydrogen fires. Int J Hydrogen Energy 36(3):2367–2373
15. Hjelmgren J (2002) Dynamic measurement of pressure—a literature survey. SP Technical Research Institute of Sweden

Chapter 3
Flame Behavior of Jet Diffusion Fire

Contents

3.1 Jet Diffusion Flame in Still Air 39
    3.1.1 Flame Length.. 39
    3.1.2 Lift-Off Distance... 41
    3.1.3 Radiative Fraction 43
    3.1.4 Application of Flame Length, Lift-Off Distance and Radiative Fraction
        Correlations 47
3.2 Jet Diffusion Flame in a Complex Boundary 53
    3.2.1 Effect of Cross Wind 53
    3.2.2 Effect of the Leakage Exit Shape 54
    3.2.3 Effect of Obstacle or Impinging Jet Flame............................ 56
    3.2.4 Effect of the Solid Particle 58
    3.2.5 Effect of the Pit 59
    3.2.6 Effect of the Underwater 61
3.3 Summary 62
References 62

The ignition following a high-pressure gas leakage can lead to a large jet diffusion fire. The intensive combustion of jet fires can release a significant amount of thermal energy, potentially causing loss of life and property damage. Therefore, it is of great interest to investigate the combustion dynamics of jet diffusion fires. In general, the flame behavior of jet fire is dominated by the competition between the exit momentum and the flame buoyancy [1]. As the exit velocity decreases during the transient leakage, the importance of the exit momentum diminishes, while the role of the flame buoyancy increases, driving the evolution of the jet flame behavior. The evolution can be quantified by the variations of the lift-off distance, flame height and radiative fraction of the jet fire.

In addition to the exit momentum and the flame buoyancy, boundary conditions can also significantly affect the behavior of a jet diffusion flame. As shown in Fig. 3.1, numerous potential boundary conditions could affect a jet diffusion fire in a hypothetical accident scenario. Studies have examined jet diffusion fires in confined spaces

---

© The Author(s), under exclusive license to Springer Nature Singapore Pte Ltd. 2024
K. Zhou, *Jet Fire Due to Gas Leakage*, Springer Series in Reliability Engineering,
https://doi.org/10.1007/978-981-97-5329-1_3

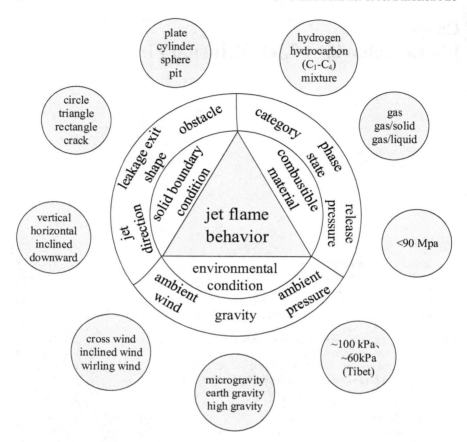

Fig. 3.1 Different boundary conditions that potentially affecting the jet diffusion flame

[2], under the low- or high-pressure atmospheric conditions [3, 4], in the presence of cross wind [5, 6], within a vortex flow field [7], in an explosive crater [8], or ejected from the leakage exits of various shapes [9]. Furthermore, interactions between jet diffusion flames and obstacles [10], liquid particles [11] or solid particles [12], as well as multi jet diffusion flames [13] have also been intensively studied.

Leaks and fires in process industry are the focus of this work. Gas storage vessels and transmission pipelines can rupture, resulting in leakage exits of various shapes. The jet flame would impinge the surrounding pipes and vessels during its development. A leakage due to the rupture of underground pipelines could entrain the nearby soil or sand particles. An explosion of underground pipelines often causes a big crater, within which a large jet fire may arise following the rupture. In addition, underwater pipelines can leak, causing flames to burn above the water surface. Therefore, in this chapter, a comprehensive review of jet fires in still air is presented as a baseline, followed by an exploration of the interactions between the jet flame

3.1 Jet Diffusion Flame in Still Air

3.1.1 Flame Length

The flame length is an important parameter to quantify the jet flame behavior. In the early studies of Hottel and Hawthorne [14, 15], the jet flame behavior transitions from laminar to turbulent, as the leakage exit velocity increases. The mixing between fuel and air streams is dominated by the molecular diffusion for the laminar jet flame, while it is the eddy diffusion for the turbulent jet flame. The rate of molecular diffusion is much lower than that of eddy diffusion, which explains that the flame length firstly increases, then slightly decreases and finally keeps constant as the exit velocity increases. Becker and Liang [16] suggested the Richardson number to well collapse the flame length, and four different correlations are proposed in four different ranges of Reynolds number and Richardson number, respectively. Lately, Kalghatgi [16] intensively examined the correlation of Becker and Liang, using the data of hydrogen, propane, methane and ethylene jet fires in subsonic and choked leakage flows. Comparison shows that the correlation of Becker and Liang cannot fit the choked jet fire.

In addition to the Richardson number, the Froude number at the leakage exit was intensively used to fit the dimensionless jet flame length normalized by the exit diameter [17, 18], and the Reynolds number at the leakage exit was also applied to collapse the dimensionless flame length of jet fire [19]. However, the data fitting gives the coefficients depending on the gas type, in terms of Froude number or Reynolds number, as shown in Table 3.1. Therefore, the flame Froude number that couples the Froude number at the exit and the physicochemical properties of fuel gas, was developed to fit the jet flame length of different fuel gases [20]. The flame length correlation can be written as

$$L^* = \begin{cases} \frac{13.5Fr_f^{2/5}}{(1+0.07Fr_f^2)^{1/5}} & Fr_f < 5 \\ 23 & Fr_f \geq 5 \end{cases} \tag{3.1}$$

in which the dimensionless flame length (L^*) and the flame Froude number (Fr_f) can be expressed by, respectively,

$$L^* = \frac{Lf_s}{d_2(\rho_2/\rho_a)^{0.5}} \tag{3.2}$$

$$Fr_f = Fr^{1/2}(\rho_2/\rho_a)^{-1/4}f_s^{3/2}(\Delta T_f/T_a)^{-1/2}, \quad Fr = u_2^2/gd_2 \tag{3.3}$$

Table 3.1 Collection of formula on flame length and lift-off height of jet diffusion fires

Author and ref.	Fuel	Orientation	d_2/mm	Fr	L/d_2	S/d_2
Suris et al. [17]	Hydrogen	Vertical	1.5–11	$Fr \leq 3 \times 10^4$	$L/d_2 = (14 \text{ or } 16)Fr^{0.2}$	–
	Methane				$L/d_2 = (27 \text{ or } 29)Fr^{0.2}$	
	Propane				$L/d_2 = 40Fr^{0.2}$	
Sonju et al. [18]	Methane	Vertical	2.3–5	$Fr \leq 1 \times 10^5$	$L/d_2 = 21Fr^{0.2}$	$S/d_2 = 3.6 \times 10^{-3}(u_2/d_2)$
	Propane		2.3–80		$L/d_2 = 27Fr^{0.2}$	
Costa et al. [23]	Methane	Vertical	5–8	$Fr \leq 1 \times 10^4$	$L/d_2 = 25Fr^{0.2}$	$S/d_2 = 3.1 \times 10^{-3}(u_2/d_2)$
Santos et al. [24]	Ethylene	Vertical	5–8	$Fr \leq 2 \times 10^4$	$L/d_2 = 24Fr^{0.2}$	$S/d_2 = 0.8 \times 10^{-3}(u_2/d_2)$
	Propane				$L/d_2 = 36Fr^{0.2}$	$S/d_2 = 2.6 \times 10^{-3}(u_2/d_2)$
Kiran et al. [25]	LPG	Vertical	2.2	$Fr \leq 4.5 \times 10^4$	$L/d_2 = 30Fr^{0.2}$	$S/d_2 = 1.8 \times 10^{-3}(u_2/d_2)$
Palacios et al. [19]	Propane	Vertical	10–43	$Fr \leq 5 \times 10^5$	$L/d_2 = 61Fr^{0.11}$	$S/d_2 = 0.62Fr^{0.3}$
Gopalaswami et al. [26]	LPG	Horizontal	19	$Fr \leq 2 \times 10^5$	$L/d_2 = 23Fr^{0.2}$	$S/d_2 = 9.7 \times 10^{-3}(u_2/d_2)$
Palacios et al. [19]	Propane	Vertical	10–43	$Re \leq 3 \times 10^6$	$L/d_2 = 5.8Re^{0.27}$	$S = 6 \times 10^{-4}Re^{0.5}$

3.1 Jet Diffusion Flame in Still Air

where f_s is the mass fraction of fuel gas at stoichiometric conditions, ΔT_f is the adiabatic flame temperature (T_{ad}) rise above the ambient temperature (T_a), and g is the acceleration of gravity. Equation (3.1) is developed to initially predict the subsonic jet flame [20], and then is validated by the choked or supersonic hydrogen jet flame [21]. In use of Eq. (3.1) for the prediction of choked jet fires, the exit diameter (d_2), velocity (u_2) and density (ρ_2) in Eqs. (3.2) and (3.3) should be replaced by those $(d_3, u_3$ and $\rho_3)$ at the Mach disc (see Fig. 2.1). Notice the variation of the leaked gas density with the time during the high-pressure gas transient leakage.

However, Bradley et al. [22] argued that the jet flame structure in a choked flow regime is different from that of the subsonic jet flame. A dimensionless flow number (U^*) proposed for the choked and subsonic flow, is found to well fit the choked and subsonic jet flame height. A vast experimental database covering 880 flame heights, is used to fit the correlation. The dimensionless flame height can be expressed by

$$
L/d_2 = \begin{cases}
81U^{*0.46} & \text{for } U^* < 10, \text{ in the buoyancy and turbulent} \\
& \quad \text{subsonic regime} \\
230 & \text{for } 10 < U^* < 80, \text{ in the transition regime} \\
42U^{*0.4} & \text{for } U^* > 80, \text{ in the choked and turbulent} \\
& \quad \text{supersonic regime}
\end{cases} \tag{3.4}
$$

in which $U^* = (u_2/u_L)(d_2u_L/\upsilon_g)^{-0.4}(p_1/p_a)$ and $U^* = (u_3/u_L)(d_3u_L/\upsilon_g)^{-0.4}(p_1/p_a)$ in the subsonic and supersonic regimes, respectively. Notice that the leakage exit diameter (d_2) is used as the characteristic length in the dimensionless flame length, despite the flow regime. Here, u_L and υ_g are the maximum laminar burning velocity and the kinematic viscosity of the fuel gas, respectively. The u_L equals 3.07 m/s, 0.39 m/s and 0.45 m/s for H_2, CH_4 and C_3H_8, respectively. Notice the variation of the gas kinematic viscosity of the gas temperature, and the variation of gas temperature and pressure with the time during the high-pressure gas transient leakage.

Both Eqs. (3.1) and (3.4) indicate a special regime in which the flame length height is independent on the gas leakage flow velocity. However, Eq. (3.4) tells one more regime in which the flame length height continues to increase after the leakage flow is choked and supersonic.

3.1.2 Lift-Off Distance

The base of jet flame begins to detach the leakage exit, as the exit velocity is large enough. The lift-off distance (S) normalized by the exit diameter is generally correlated by the global strain rate at the leakage exit. The correlation of the lift-off distance can be written as

$$
S/d_2 = C(u_2/d_2) \tag{3.5}
$$

in which C is a coefficient determined by data fitting. $C = 2.65 \times 10^{-5}$ s for the vertical hydrogen jet fire [27], $C = 3.60 \times 10^{-3}$ s for the vertical methane jet fire [18], $C = 2.13 \times 10^{-3}$ s for the vertical propane jet fire [28], and $C = 9.55 \times 10^{-3}$ s for the horizontal propane jet fire [28]. Table 3.1 shows the different coefficients in the available literature. The limitation of Eq. (3.5) is how to determine the coefficient using the physicochemical properties of fuel gas.

The lift-off phenomenon of a jet diffusion flame is generally explained by the model of premixed flame propagation [29, 30]. The base of jet flame should stabilize at a position where the local gas flow velocity equals the flame burning speed of premixed fuel and air mixture. The dynamic viscosity and density of fuel gas would affect the decay rate of gas jet flow velocity. Obviously, the flame burning speed, dynamic viscosity and density of fuel gas should be coupled into the lift-off distance correlation. Kalghatgi [30] identified dimensionless groupings of the various flow and gas parameters that affect the lift-off distance, and proposed

$$\frac{Su_L}{v_g} = C\frac{u_2}{u_L}\left(\rho_2/\rho_a\right)^{1.5} \tag{3.6}$$

In Eq. (3.6), the dimensionless coefficient C of approximately 50 is determined by data fitting of hydrogen, methane, propane and ethylene jet flames [30]. In particular, the data for fitting covers the choked and supersonic hydrogen jet flames, in which the exit velocity (u_2) and density (ρ_2) in Eq. (3.6) are replaced by those (u_3 and ρ_3) at the Mach disc (see Fig. 2.1). However, the power to the density ratio in Eq. (3.6) decreased from 1.5 to 1.0, after further studies by Wu et al. [31, 32]. Lately, Bradley et al. [22] used a vast experimental data base to reanalyze Eq. (3.6), but with the power of 1 to the density ratio, and gave the correlation of the lift-off distance by

$$\frac{Su_L}{v_g} = 52.4\left[\frac{u_2}{u_L}\left(\rho_2/\rho_a\right) - 11.8\right]^{1.04} \tag{3.7}$$

As compared to Eq. (3.6), Eq. (3.7) tells the fact that the lift off does not appear until the exit velocity is over a critical value.

The dimensionless flow number was also used to fit the lift-off distance in the subsonic and supersonic regimes [22]. The optimal correlation of the lift-off distance is given by

$$(S/d_2)f = 0.11U^* - 0.2, \text{ in the subsonic regime}$$
$$(S/d_2)f^{0.2} = -54 + 17\ln\left(U^* - 23\right), \text{ in the choked and supersonic regime} \tag{3.8}$$

The quantitative definition of U^* in Eq. (3.8) refers to Eq. (3.4). Here f is the ratio of fuel to air moles in fuel–air mixture for the maximum burning velocity (u_L).

3.1 Jet Diffusion Flame in Still Air   43

3.1.3 Radiative Fraction

The radiative fraction (χ_r) is defined as the ratio of the total radiant energy escaping from the jet flame ($\dot{Q}_r$) to the total heat release rate ($\dot{Q}$), i.e.

$$\chi_r = \dot{Q}_r / \dot{Q} \tag{3.9}$$

It is an important parameter to evaluate the potential threaten or damage due to the radiant heat from jet flames. The radiative fraction of natural gas jet fires was exponentially correlated with the leakage exit velocity in the experimental work of Chamberlain [33]. Markstein and de Ris [34] measured the radiative fractions of jet fires of four different hydrocarbon fuels, and fitted them against the total heat release rate by means of power function, respectively. The global residence time, derived from the convective timescale, was also used to correlate the radiative fractions of methane, ethylene and propane jet fires, respectively [35], and hydrogen jet fires [21]. The flame length was also plotted against the radiative fraction of large hydrogen jet fires [36]. Obviously, these correlations of radiative fraction are so empirical that they strongly depend on the test conditions.

It is of great interest to detail the radiative fraction correlation that couples the global residence time and the correction factor based on the differences in thermal emittance of combustion gases of different fuels. Molina et al. [37] determined the correlation as Eq. (3.10), using the radiative fraction data of large hydrogen, methane, and carbon monoxide/hydrogen mixture jet fires.

$$\chi_r = 0.085 \log\left(\tau_G a_p T_{ad}^4\right) - 1.16 \tag{3.10}$$

in which τ_G is the global residence time, a_p is the Planck-mean absorption coefficient, and T_{ad} is the adiabatic flame temperature. $\tau_G = \rho_f W_f^2 L_f f_s / 3\rho_2 d_2^2 u_2$ in which ρ_f is the density of the mixture of the products of a stoichiometric flame, and W_f and L_f are the visible flame width and length, respectively. Notice that the unit of $\tau_G a_p T_{ad}^4$ is $ms \cdot K^4 \cdot m^{-1}$. The visible flame length is the flame length minus the lift-off distance, i.e. $L_f = L-S$. A lot of studies measured the ratio of the flame width to length, and the ratio is approximately 0.17 [35, 38].

Zhou et al. [39] conducted a theoretical analysis to clarify the key parameters that dominate the radiative fraction of jet fires. They proposed a completely new dimensionless relationship to couple the $\chi_r, f_s, \rho_2 / \rho_a$ and Fr_f, expressed by

$$\frac{\chi_r f_s}{\sqrt{\rho_2 / \rho_a}} \propto Fr_f \tag{3.11}$$

Obviously, Eq. (3.11) can be reducible to be a simple form of $\chi_r \propto Fr^{1/2}$, by which the Froude number at the leakage exit is also used to correlate the radiative fraction of propane jet fires [28]. In order to evaluate the capacity of Fr and Fr_f to fit the radiative fraction, data were extracted from 8 publications that cover the works on the hydrogen, methane and propane jet fires [39]. Table 3.2 summarizes the data bank that involves the Froude number at the leakage exit, the radiative fraction, and the fuel property parameters.

Figure 3.2 presents the radiative fraction versus the Froude number at the leakage exit, for hydrogen, methane and propane jet fires. The radiative fraction firstly keeps constant and then decreases significantly, as the Froude number increases, for all three types of fuel. The decrease of radiative fraction relates to the appearance of bright blue flame at the lower portion, for the hydrocarbon jet fires, while it could result from much air entrained from the lower portion that cools the hydrogen jet flame. Anyway, the turning point for the radiative fraction decreasing can help to define a critical Froude number. The critical Froude number is approximately 500, 30 and 10 for the hydrogen, methane and propane jet fires, respectively. Below the critical Froude number, the radiative fraction is approximately 0.1, 0.2 and 0.3 for the three jet fires in still air, respectively. The absence of CO_2 radiation bands reduces the radiative fraction of hydrogen jet flames as compared to flames with CO_2 as a product. Also the sooting propane jet flames apparently hold larger radiative fraction than those of non-sooting methane jet flames.

In addition, some other factors that affect the radiative fraction can be clarified and evaluated. As shown in Fig. 3.2a, the radiative fraction seems to increase, as the Froude number is over 4000. In detail, the leakage exit velocity (over 1000 m/s) is over the local sound speed, and the normal shock near the leakage exit of the choked jet flow holds a positive effect on the radiative fraction. The normal shock imposes a restraint on the air entrainment and mixing rate in the flame lift-off region [46], and thus improves the flame radiation temperature in the downstream combustion region. As indicated in Fig. 3.2b, the background wind significantly increases the radiative fraction for the little Froude number. However, as the Froude number increases, the effect of background wind gradually reduces and the radiative fraction decays to those of jet fires in still air. The difference relates to the variation of air entrainment or air–fuel mixing with the ratio of the crosswind velocity to jet velocity. As presented in Fig. 3.2c, the jet direction holds little effect on the radiative fraction.

In short, the fuel type, normal shock, crosswind momentum flux, jet momentum flux and flame buoyancy that account for differences in the radiative characteristics, should be considered to develop a generalized correlation of radiative fraction for jet fires. Accordingly, Eq. (3.11), together with the fuel property parameters in Table 3.2, is used to replot all the radiative fraction data in still air in Fig. 3.2. As shown in Fig. 3.3, all the measured radiative fraction collapse onto one curve when plotted against the flame Froude number. In detail, the radiative fraction almost keeps constant in the buoyancy-controlled regime ($Fr_f < 0.1$), and then decreases in the transition regime ($0.1 < Fr_f < 10$), and finally increases in the momentum-controlled regime ($Fr_f > 10$). The first turning point results from the shrinkage of upper yellow soot region and the enlargement of the lower blue flame part, while the second turning point is due to the effect of the normal shock. There is no doubt that the fitting correlation can be used to estimate the radiative fraction of jet fires of hydrogen, hydrocarbon or their mixture.

Table 3.2 Parameter ranges and fuel types for the data fitting of radiative fraction [39]

Fuel	d_2 (mm)	u_2 (m/s)	$\dot{m}_2$ (kg/s)	$Fr^{1/2}$	χ_r	f_s	ρ_2/ρ_a	T_{ad}	c_p^a (kJ/kg K)	ρ_r^a (kg/m³)	$\rho_r c_p \Delta T_r$ (kJ/kg)
H$_2$ [21]	1.91, 5.08, 7.94	88–1233	2.1 × 10^{-5}–0.359	641–5109	0.048–0.089	0.0283	0.069	2382	2.72	0.122	595
H$_2$ [40]	154, 250, 686	26–983	1.9–126	10–533	0.086–0.153						
CH$_4$ [41]	107–600	22–555	–	9–434	0.121–0.322	0.0551	0.552	2226	2.19	0.148	529
CH$_4$ [42]	5	4–62	–	20–279	0.136–0.195						
C$_3$H$_8$ [42]	5	1–57	–	5–259	0.155–0.376	0.0602	1.517	2267	2.27	0.149	565
C$_3$H$_8$ [43]	2.06–250	–	–	10^{-3}–360	0.113–0.358						
C$_3$H$_8$ [24]	5, 6, 8	5–48	–	18–215	0.103–0.222						
C$_3$H$_8$ [44]	1	9–57	–	93–580	0.139–0.171						
C$_3$H$_8$ [45]	19.1	28–206	0.015–0.110	65–476	0.050–0.240						

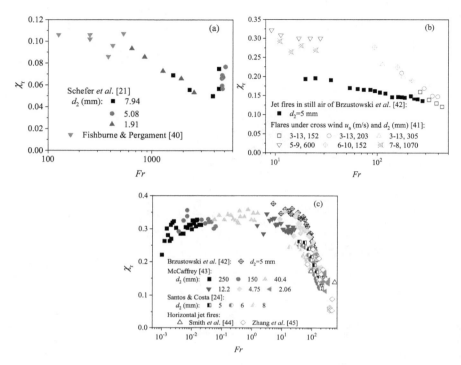

Fig. 3.2 Radiative fraction versus Froude number at the leakage exit: **a** hydrogen jet fires in still air, **b** methane jet fires, and **c** propane jet fires in still air [39]. Jets are classified as flares as they are used for the safe disposal of unwanted flammable gases from pipes with orifice exit diameters over 150 mm. The radiative fraction from Santos and Costa [24] has been divided by the correction factor $C^* = 0.85$

Fig. 3.3 Radiative fraction versus flame Froude number for hydrogen and hydrocarbon jet fires in still air [39]. In the Figure, I, II and III stand for the buoyancy-controlled regime, transition regime and momentum-controlled regime, respectively. The linear and power function fittings are conducted in regime I and the other two regimes, respectively, to roughly show the form of Eq. (3.11)

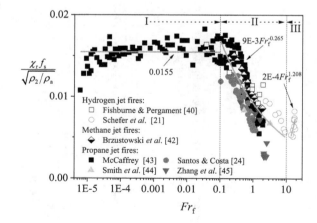

3.1 Jet Diffusion Flame in Still Air   47

3.1.4 Application of Flame Length, Lift-Off Distance and Radiative Fraction Correlations

In the experimental test of Proust et al. [36], besides the gas pressure, temperature and mass flow rate as detailed in Sect. 2.6, the flame length and the radiant heat flux were recorded by a video camera at about 5 m away from the centerline and CAPTEC flux meters at five positions, respectively. Note that their hydrogen jet fire tests were done in still air. Accordingly, the measured flame length is used to test the correlations of flame behavior in Sects. 3.1.

As concluded in Sect. 2.6, the van der Waals gas leakage model holds a stronger robustness than the other two models. Accordingly, the state properties and flow parameters calculated by the van der Waals gas leakage model and the notional nozzle model, are input into the correlations of flame length. The flame length correlations based on Fr number, Fr_f number, and U^* number require the known flow velocity and diameter at the Mach disc (u_3 and d_3) and the leakage exit (u_2 and d_2), for the choked and subsonic flows, respectively. In addition, the Fr_f number requires the density ratio of the fuel gas to the surrounding air, while the U^* number needs the gas pressure in vessel and the kinematic viscosity of the fuel gas.

Figure 3.4 shows the required parameters predicted by the van der Waals gas leakage model and the notional nozzle model, for $d_2 = 2$ mm. As shown, these parameters significantly vary with time. The calculation of the kinematic viscosity uses the polynomial formula of $v_g = -7.50 + 1.92 \times 10^{-1}T + 1.36 \times 10^{-3}T^2 - 6.67 \times 10^{-7}T^3$ for the hydrogen in the temperature of 40–240 K. Notice $k = 1.4$ for the leakage process. The initial condition parameters input into the leakage model are listed in Table 2.3 of Sect. 2.6.

The flame length can be predicted by the three correlations based on Fr number (Table 3.1), Fr_f number (Eqs. (3.1)–(3.3)), and U^* number (Eq. (3.4)), with the parameters in Fig. 3.4. Figure 3.5 shows the comparison of flame length between the correlation calculations and the experimental measurement. The Fr number correlation significantly overestimates the flame length, even though the choice from Table 3.1 is $L/d_2 = 14Fr^{0.2}$. The Fr_f number correlation well fellows the measurement in the initial leakage period (duration < 20 s), and then gives a slight overestimation in the latter period. The U^* number correlation slightly underestimates the measured flame length. The comparison is not to show which correlation is best independently, but to tell which one could be well coupled and work with the leakage model and the notional nozzle model. In the Fr number correlation, the coefficient could be affected by the state and thermophysical properties that vary with time during the hydrogen transient release, which causes the significant overestimation. The density ratio of the leaked hydrogen to the ambient air is considered in the Fr_f number correlation, while the U^* number correlation takes into account the pressure and the kinematic viscosity. Accordingly, the accuracy of density ratio, predicted by the leakage model, would affect the calculation of the Fr_f number correlation, while the accuracy of pressure and viscosity does that of the U^* number correlation. Note the coupled effect of two uncertain parameters on the latter correlation.

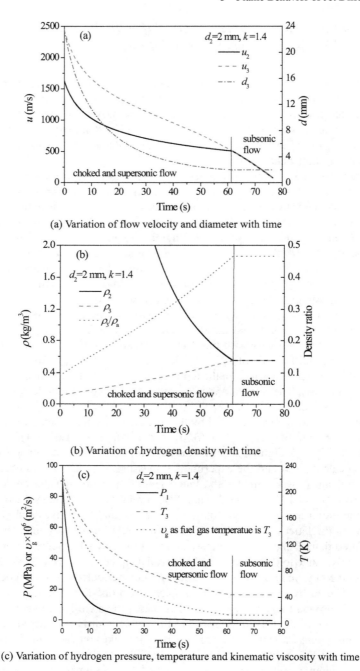

Fig. 3.4 The parameters required for the flame length correlations. They are predicted by the van der Waals gas leakage model using $k = 1.4$ and the notional nozzle model, for $d_2 = 2$ mm

3.1 Jet Diffusion Flame in Still Air

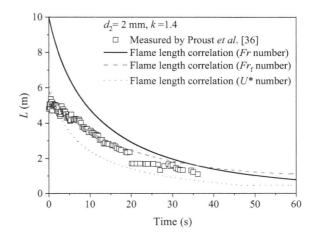

Fig. 3.5 The prediction of the flame length by three correlations against the measurement ($d_2 = 2$ mm, $k = 1.4$)

The assumption of the isentropic leakage process, i.e., $k = 1.4$ for hydrogen, could be not good enough to fit the experimental test of Proust et al. [36], as compared to $k = 1.1$, as discussed in Sect. 2.6 The model prediction of the flow parameter, state and thermophysical property, and flame length, with $k = 1.1$, is presented in Fig. 3.6. The calculation of the kinematic viscosity uses the polynomial formula of $v_g = -43.5 + 7.56 \times 10^{-1}T - 1.52 \times 10^{-3}T^2 + 4.13 \times 10^{-6}T^3$ for the hydrogen in the temperature of 150–300 K. As shown in Fig. 3.6d, the Fr number correlation still largely overestimates the flame length, while the Fr_f and U^* number correlations seems to give a good prediction. Accordingly, the flame length versus the hydrogen pressure in vessel is used to further test the leakage model and the flame correlations based on Fr_f and U^* numbers.

Figure 3.7 shows the comparison of the flame length versus the hydrogen pressure in vessel between the model prediction and the experimental measurement, for the leakage exits of $d_2 = 1$ mm, 2 mm and 3 mm. The van der Waals gas leakage model can predict the variation of the hydrogen pressure with time, as shown in Fig. 3.6c. The Fr_f and U^* number correlations help to calculate the flame length, respectively. As presented in Fig. 3.7, the Fr_f number correlation (Eqs. (3.1)–(3.3)) shows a better prediction than the U^* number correlation (Eq. (3.4)). Thus, Eqs. (3.1)–(3.3) will be used for the following calculation and discussion on the radiative fraction and total radiant output rate.

The correlations based on the global residence time (Eq. (3.10) and the Fr_f number (Eq. (3.11)) are used to calculate the radiative fraction of the transient high-pressure jet flame. In addition to the flow and thermophysical parameters by the van der Waals leakage model with $k = 1.1$, and the flame length by Eqs. (3.1)–(3.3), the calculation of the radiative fraction also needs the lift-off distance. Here, Eq. (3.5) helps to calculate the lift-off distance of the hydrogen jet fire. In addition, the product of the mass leakage rate (Fig. 2.14a) and the combustion heat (142 MJ/kg for hydrogen) is the total heat release rate, and the product of the radiative fraction and the heat release rate is the total radiant output rate. Figure 3.8 shows the variation of the predicted

Fig. 3.6 The prediction of the flow parameter, state and thermophysical property, and flame length ($d_2 = 2$ mm, $k = 1.1$)

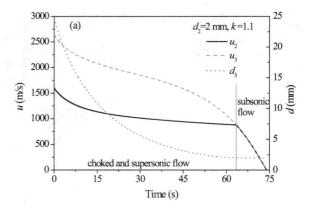

(a) Variation of flow velocity and diameter with time

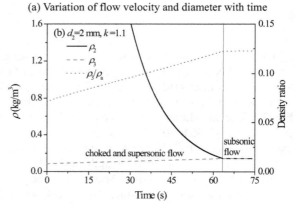

(b) Variation of hydrogen density with time

radiative fraction and total radiant output rate with the time. The radiative fraction shows a non-monotonic variation with the time, while the total radiant output rate decreases monotonically with the time. Notice the Fr_f number range of 0.10–28.48 and the $\tau_G a_p T_{ad}^4$ range of 5.67×10^{13}–3.36×10^{14} in the calculation for Fig. 3.8 a, as compared to that of 10^{-3}–18.34 in Fig. 3.3 and that of 10^{14}–5×10^{15} in the development of Eq. (3.10) from [37], respectively. Anyway, the validation of the predicted radiative property will be evaluated in Sect. 4.3 in which the radiative fraction and radiant output rate are input into the thermal radiation model against the measured radiant heat flux.

3.1 Jet Diffusion Flame in Still Air

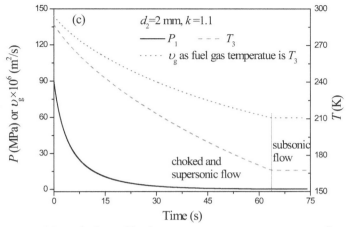

(c) Variation of hydrogen pressure, temperature and kinematic viscosity with time

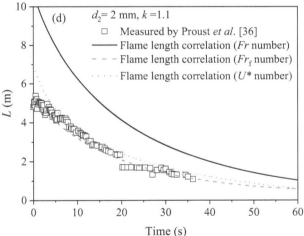

(d) Comparison of flame length versus time between prediction and measurement

Fig. 3.6 (continued)

Fig. 3.7 The prediction of the flame length (by the Fr_f and U^* correlations) versus the hydrogen pressure in vessel (by the van der Waals gas leakage model) against the measurement ($k = 1.1$)

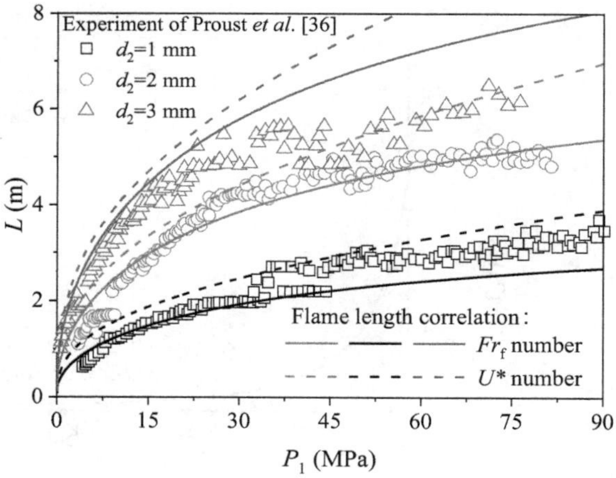

Fig. 3.8 The prediction of **a** the radiative fraction and **b** the total radiant output rate ($d_2 = 2$ mm, $k = 1.1$)

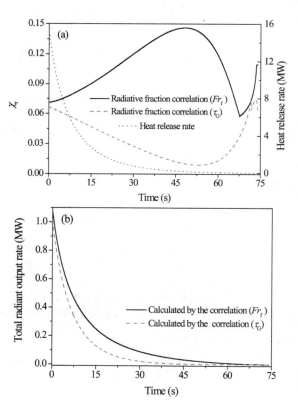

3.2 Jet Diffusion Flame in a Complex Boundary

3.2.1 Effect of Cross Wind

In cross wind, there is a shear force impacting on the jet flame. The flame buoyancy, the inertia force at the exit and the shear force jointly affects the jet flame behavior. In general, two dimensionless parameters, i.e., the Froude number at the exit (Fr) and the jet-to-crossflow momentum flux ratio (R_M) are often used to quantify the jet flame geometry in cross wind. The definition of Fr is shown in Eq. (3.3). The R_M can be expressed by $R_M = (\rho_2 u_2^2)/(\rho_a u_a^2)$ and $R_M = (\rho_3 u_3^2)/(\rho_a u_a^2)$ in the subsonic and choked flow regimes, respectively, where u_a is the velocity of transverse air stream or cross wind.

Take for example the vertical jet flame in cross wind. The down-wash and reverse flow regions form in the wake of the burner and the jet, which dominates the flame stability. As the R_M increases, the characteristic flow quantified by the streamlines, vorticity distributions and flame appearance, shows the down-wash ($R_M < 0.1$), cross-flow dominated ($0.1 < R_M < 1.6$), transitional ($1.6 < R_M < 3.0$), jet dominated ($3.0 < R_M < 10$), and strong jet ($R_M > 10$) modes in turn [47]. A further increase of R_M results in the blow off of a jet flame. The flame length increases in the cross-flow dominated mode, first decreases and then increases in the transitional mode, and increases in the jet-dominated mode, as the R_M increases. The tilt angle (θ) of the jet flame away from the vertical direction is jointly dominated by the R_M and Fr. The flame tilting behavior holds three different dominated regimes, i.e., crossflow-dominated ($R_M < 0.01$, $Fr < 0.1$), transitional ($0.10 < R_M < 10$, $0.1 < Fr < 10^3$) and low Mach number jet-dominated ($R_M > 10$, $10^3 < Fr < 10^5$) regimes [48]. In the three regimes, the decreases of θ in degree with the increase of R_M can be expressed by

$$\theta = \begin{cases} 74.4 R_M^{-0.006} \\ 61.2 R_M^{-0.050} \\ 73.2 R_M^{-0.115} \end{cases} \tag{3.12}$$

In the first regime, most of the jet fuel is entrained into the down-wash area to support combustion; thus, the flame deflects through a large tilt angle. The vortex in the wake of the burner moves upward and becomes small in the transitional regime, which causes a medium flame tilt angle. In the last regime, the reverse flow in the wake of both the nozzle and the jet results in a small flame tilt angle [48].

The blow-off of jet flame is dominated by the leakage velocity of fuel gas at the exit and the cross-wind speed. There is a limiting value of cross-wind speed beyond which a stable jet flame is not possible for a given burner and fuel gas. For cross winds below the limit, there are frequently two blow-off limits in terms of the leakage velocity of fuel gas. However, its lower and upper stability limits are much less and larger than that in the absence of cross wind [49]. The stability of the jet flame in the wind is also significantly affected by the orientation of the burner to the wind direction [50].

3.2.2 Effect of the Leakage Exit Shape

The leakage exit shape significantly affects the flame length and the lift-off distance of vertical jet fires. The flame length decreases, as the aspect ratio increases, for the rectangular vertical jet fires in both buoyancy-controlled [51] and momentum-controlled [52] modes. The flame length correlations with different fitting coefficients are proposed for the rectangular jet fires controlled by buoyancy and momentum, respectively [51, 52]. The lift-off distance also decreases with an increase of the aspect ratio of rectangular exits, for the vertical jet fire, and data fitting gives the correlation of the lift-off distance [51]. The reduction of the lift-off distance results from the asymmetry of the leakage exit, which can be further clarified from the comparison between vertical jet fires of circular, triangular, rectangular and square leakage exits [53]. The asymmetric exit increases the air entrainment and thereby improves mixing, which in turn could decrease the flame length and the lift-off distance.

The shape of leakage exit also affects the flame geometry of horizontal jet fires. As compared to the flame length and lift off for the vertical jet fires, the horizontally projected length, vertically projected height and lift-off distance are often used to quantify the flame geometry of horizontal jet fires [9], as shown in Fig. 3.9.

Figure 3.10 shows the variation of the horizontally projected flame length with the exit velocity, for the jet fires of different exit shapes. The flame length increases to reach a constant as the exit velocity increases to be over a critical value. As shown, the flame length decreases as the aspect ratio of rectangular exit increases. In addition, the circular exit produces the largest flame length of jet fire. The classical correlation can well fellow the jet flame length of circular exit, for the empirical correlation is built by data fitting of circular jet fires. Accordingly, it is of great interest to couple the empirical correlation with the exit shape coefficient. The ratio of the hydraulic

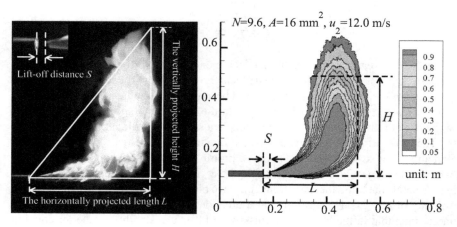

Fig. 3.9 Typical horizontal jet flame (front elevation view) and the definition of flame geometry [9]

3.2 Jet Diffusion Flame in a Complex Boundary

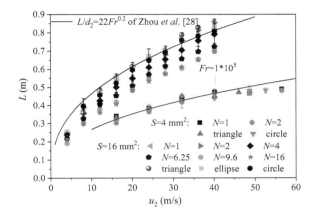

Fig. 3.10 Horizontally projected flame length versus exit velocity: experimental measurement and classical correlation calculation [9]. A is the exit area, and N is the aspect ratio of rectangular exit, and $d_2 = \sqrt{4A/\pi}$ is the equivalent diameter. The error bars indicate the test repeatability

diameter to the equivalent diameter can help to quantify the exit shape coefficient. Data fitting of vertical and horizontal, methane and propane jet fires with different exit shapes, shows that the product of the exit shape coefficient and molar stoichiometric air to fuel ratio is proportional to the phenomenological constant related to the air entrainment, and an universal correlation is proposed for the flame length [9]. In recent, the flame length correlation was validated to fit the horizontal jet fires of different crack exits [54].

Figure 3.11 shows that the vertically projected flame height firstly increases and then decreases, as the exit velocity increases, especially for the exit area of 16 mm^2, for the jet fires of different exit shapes. The flame buoyancy positively dominates the flame height at a small exit velocity, and the positive effect of flame buoyancy enhances as the exit velocity increases. However, as the exit velocity increases to be large enough, the reductive effect of exit momentum increases on the flame height. Accordingly, the Richardson number representing the ratio of the flame buoyancy to the exit momentum, can well fit the ratio of the projected flame height to length of the horizontal jet fires of different exit shapes [9]. In recent, the Richardson number was validated to fit the flame height of the horizontal jet fires ejected from different crack exits [54].

Figure 3.12 shows the variation of the lift-off distance with the exit velocity for jet flames ejected from exits of different shapes. The lift-off distance seems to linearly increase as the exit velocity increases, for the constant exit area and shape. The exit shape significantly affects the slope of linear variation. In use of the Eq. (3.8) to fit the lift-off distance, the hydraulic diameter instead of the equivalent diameter, is suggested as the characteristic diameter in dimensionless flow number and dimensionless lift-off distance [9].

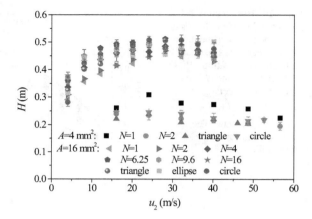

Fig. 3.11 Vertically projected flame height versus exit velocity under different exit shapes [9]. The error bars show the test repeatability

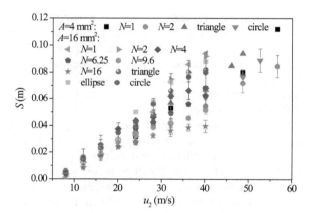

Fig. 3.12 Lift-off distance versus exit velocity under different exit shapes [9]

3.2.3 Effect of Obstacle or Impinging Jet Flame

The obstacle can change the jet flow direction and also slow down the flow velocity, which complexes the combustion dynamics of impinging jet flame. Figure 3.13 shows the evolution of flame pattern as the leakage exit velocity increases when the jet flow impinges on a plate. The color of propane jet flame varies from brightly yellow to transparently blue. When the exit velocity is little, less air is entrained before the impingement and the impinging jet flame is diffusive. However, the air entrained into the propane jet flow gradually increases as the exit velocity increases, and the air-propane mixture jet impinges the plate to form the premixed combustion along the plate. In particular, the flow velocity of air-propane mixture at the impinging point is larger than the flame burning velocity, a circular hole of no flame appears around the stagnation point. In short, the impinging jet flame gradually transitions from diffusive combustion to premixed combustion, which indicates the significant

3.2 Jet Diffusion Flame in a Complex Boundary

Fig. 3.13 Flame pattern versus exit velocity of horizontally impinging jet fire. The circular exit of 2 mm in diameter is 20 cm away from the plate [10]

increase of flame temperature, and the flame pattern gradually approaches to be approximately circular, as the exit velocity increases. Note that the exit velocity of 56.46 m/s is large enough to blow off the free jet flame of propane ejected from 2 mm exit, if there is no plate to slow down the gas flow and stabilize the combustion (Fig. 3.13).

Figure 3.14 shows the impingement of a horizontal propane jet flame on a hollow steel cylinder. The yellow flame also gradually becomes completely blue, and the flame is prone to wrap around the cylinder, as the leakage exit velocity increases. It is again demonstrated that the obstacle slows down the propane jet flow and enhances the air–fuel mixture before combustion around the obstacle surface. The unique behavior of impinging jet flame obviously enhances the heat transfer from the burning flame to the obstacle.

The convective heat transfer resulting from the jet flame impingement, as well as the thermal radiation from jet fires, is the escalation vector of domino effect in process industry. The calculation of the convective heat, i.e., Newton's law of cooling, needs to know the temperature profile and extension area of the impinging jet flame. Some empirical correlations are proposed to quantify the extension area [10, 55, 56] and the temperature distribution [10, 57] of the impinging jet flame. The difference of test conditions, e.g., the impinging direction and the Froude number at the exit, derives different correlations by data fitting. Comparatively speaking, a universe correlation that couples the Karlovitz stretch factor, exit diameter and exit-plate spacing is developed to predict the flame extension area. It is cited here [10]

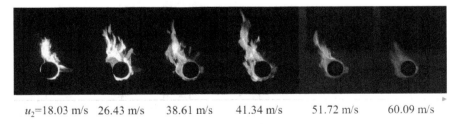

Fig. 3.14 Flame pattern versus exit velocity of horizontally impinging jet fire. The circular exit of 3 mm in diameter is 15 cm away from the pipe of 8.9 cm in diameter

$$S/d_2^2 = 265746\Lambda^{1.2} (0 \leq \Lambda \leq 0.14) \tag{3.13}$$

where $\Lambda = (u_2/u_L)(d_2 u_L/v_g)^{-0.4}(d_2/D)$ in which D is the spacing between leakage exit and the plate. Notice that Eq. (3.13) covers the data of impinging jet flame in a large range of Froude number and jet direction. For the impingement of jet flame on a curved surface, e.g., that as depicted in Fig. 3.14, Eq. (3.13) fails to predict the flame extension area around the obstacle.

3.2.4 Effect of the Solid Particle

The solid particle, e.g., sand and soil, could be entrained into the jet diffusion flame of high flow velocity. Figure 3.15 shows the typical flame behavior of propane-sand jet flow against the pure propane jet flow. The addition of sand increases the lift-off distance, but decreases the visible flame length. However, the sum of the lift-off distance and the visible flame length, i.e., the entire flame length, would increase due to the addition of sand [12]. In addition, the radiative fraction significantly decreases as the entrained sand increases [12].

The entrained sand plays a role of heat sink to decrease the flame temperature, and thus reduces the flame burning velocity. Notice the appearance of the gas-air-sand premixed combustion at the base of lifted flame of the gas-sand jet diffusion fire. In addition, the sand in the lift-off region isolates the surrounding oxygen and dilutes the fuel to inhibit the combustion. Both effects jointly result in a larger lift-off distance for jet fires with the addition of sand.

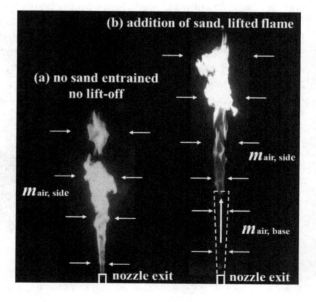

Fig. 3.15 Comparison between propane jet flame behaviors **a** without and **b** with the addition of sand. ($d_2 = 10$ mm, $m_2 = 0.401$ g/s) [12]

3.2 Jet Diffusion Flame in a Complex Boundary   59

The jet flame is more prone to produce the lift off phenomenon under the effect
of the entrained sand. Accordingly, the air entrained into the flame volume does not
only come from the lateral side of flame, but also from the base of lifted flame.
Moreover, the particle-laden flow enhances the air entrainment rate of fire plume
[58], and thus makes the fire plume more vulnerable to collapse [59]. In short, more
air can be entrained into the jet flame with the addition of sand, which reduces the
visible flame length.

The entrained sand reduces the flame temperature, which indicates the increase
of the flame Froude number in Eq. (3.3). Accordingly, the entire flame length of
buoyancy-dominated jet fire increases, as told by Eq. (3.1). At the same time, the
reduction of the flame temperature decreases the radiative fraction due to the addition
of sand.

3.2.5 Effect of the Pit

The underground pipeline is often reported to explode and eject the overlying soil with
the formation of a big pit. The pit size and shape depend on the pipeline diameter
and burial depth, operating pressure, rupture length, and type of surrounding soil
[60]. The pit would significantly affect the nature of the gas leakage in turn and then
influences the subsequent jet fire. Accordingly, it is of great interest to simulate a jet
flame in a pit.

As shown in Fig. 3.16, as the leakage velocity or mass flow rate increases, the
jet fire behavior show three different flame patterns in a pit. They are the impinging
jet flame (IJF), the transitive jet flame (TJF) and the jet flame ejected from the pit
top (JFEPT) [8]. The three different flame patterns indicate the different mode of air
entrained into the pit. The upper flame outside the pit presents a periodic pulsation,
despite the three flame patterns. However, there is an important difference of the
flame behavior inside the pit. The IJF exhibits a steady behavior within the pit, for
it is dominated by the nozzle exit momentum, and the fresh air is directly entrained
into the pit through the large gap between the pit top and the upper flame outside the
pit. In comparison, inside the pit the flames of TJF and JFEPT are quasi-steady with
a periodic phenomenon. A little flame intermittently appears and disappears inside
the pit for the JFEPT, because the fresh air is entrained into the pit by the lateral eddy
in the flame bottom. The lateral eddy is intermittent due to the flame buoyancy. As
for the TJF, a majority of fresh air can flow into the pit through the right gap, and a
minority of fresh air is entrained into the pit by the counter clockwise lateral eddy
on the left.

As the jet flame in a pit evolves from IJF to TJF and JFEPT due to the increase
of mass leakage rate, the radiative fraction of jet flame increases, which goes against
the law of the Froude number at the leakage exit in Fig. 3.2. As shown in Fig. 3.2,
as the leakage exit momentum begins to play more and more significant role in the
jet flame behavior, the radiative fraction starts to decreases. The IJF is dominated
by the leakage exit momentum, and thus its radiative fraction well follows the law

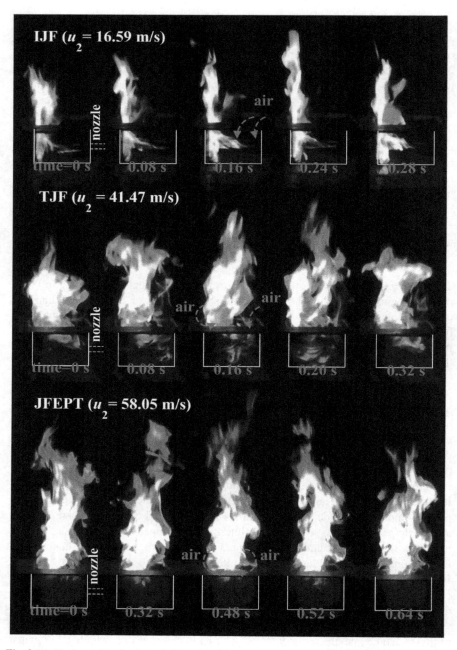

Fig. 3.16 Horizontal jet fire in a pit. The pit is 20 cm in length, 20 cm in width and 15 cm in depth [8]. The nozzle exit is 3.2 mm in diameter, i.e., $d_2 = 3.2$ mm

3.2 Jet Diffusion Flame in a Complex Boundary   61

described in Fig. 3.2. However, the JFEPT should hold an exit diameter equaling the pit top opening instead of the nozzle exit. Accordingly, the radiative fraction of JFEPT can be predicted using the Froude number at the top opening of the pit [8].

The IJF, TJF and JFEPT can be distinguished using the critical value of the dimensionless heat release rate that couples the pit size and the heat release rate of leaked gas combustion. The flame length correlation is given for the TJF and JFEPT by data fitting against the dimensionless heat release rate [8]. Accordingly, the correlation of flame length in Sect. 3.1.1 should be replaced by that in the literature [8] when the jet flame occurs in a pit.

3.2.6 Effect of the Underwater

The rapid development of the offshore oil and gas industry increases the length of the natural gas pipeline underwater. The leaked gas should first form the bubble plume underwater before reaching the water surface for combustion. The bubble plume would significantly affect the flame behavior above the water surface. For example, the upward velocity of gas is non-uniform when it leaves the water surface to enter the burning flame. Additionally, the flame base diameter varies with the time due to the fountain behavior on the water surface. These phenomena make it unique and different from the liquid pool fire on a water surface.

The water depth and gas leakage flow rate jointly dominate the behavior of the burning flame above the water surface. The ignition success indicates the gas concentration within the lower and upper limits of the flammability. Figure 3.17 shows the transient variation of the flame behavior. It could include the instantaneous burning state, violent burning state, and unstable or stable burning state. The violent burning state indicates the gas accumulation near the water surface before ignition. The unstable and stable burning states are characterized by the self-extinguishment and self-sustainment of flame, respectively. As indicated in Fig. 3.17, the increase of the gas leakage rate from 12 L/min to 14 L/min, triggers the transition from the unstable burning state to the stable burning state. For the water depth of 0.45 m, the gas leakage rate of 12 L/min is not enough to maintain the gas concentration stable at the lower flammability limit, which results in the flame extinguishment. In comparison, the 14 L/min can support the gas concentration for the self-sustaining flame. The difference could be explained by the water evaporation that can dilute the gas concentration. The water evaporation rate depends on the heat feedback from the burning flame to the water surface. The competitive mechanism between the gas leakage rate and the water evaporation rate determine whether the gas concentration could reach the lower flammability limit or not.

For the stable burning flame above the water surface, the flame base diameter increases as the water depth and gas leakage flow rate increase. Its flame pulsation frequency is less than that of liquid pool fires, and is larger than that of gas jet diffusion fires in still air. The flame height increases with the decrease of water depth and the increase of gas leakage flow rate, but it is less than those of liquid pool fires

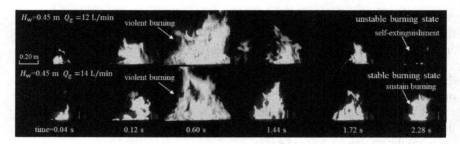

Fig. 3.17 Instantaneous images for the burning behavior of leakage gas on the water surface [61]

and gas jet diffusion fires in still air. Note lots of water vapor in the burning flame due to the water evaporation. The water vapor reduces the flame temperature, and thus diminishes the flame buoyancy, which causes the decrease of fire plume height. In addition, the water vapor also significantly affects the flame radiation property.

3.3 Summary

Different available correlations for flame length, lift-off distance and radiative fraction of jet fires in still air are reviewed. These correlations can be integrated into the gas transient leakage model discussed in Chapter 2 to calculate the geometry of the jet flame. A comparison between model calculations and experimental measurements indicates that the correlations for flame length and radiative fraction, which are based on the flame Froude number, are superior. In addition, the behavior of jet fires can be significantly affected by various boundary conditions. The effect of the cross wind, leakage exit shape, obstacle, solid particle, pit and underwater on the jet fire behavior are also reviewed.

References

1. Kessler A, Schreiber A, Wassmer C et al (2014) Ignition of hydrogen jet fires from high pressure storage. Int J Hydrog Energy 39(35):20554–20559
2. Cha MS, Chung SH (1996) Characteristics of lifted flames in nonpremixed turbulent confined jets. Symp (Int) Combust 26(1):121–128
3. Wang Q, Hu L, Zhang M et al (2014) Lift-off of jet diffusion flame in sub-atmospheric pressures: an experimental investigation and interpretation based on laminar flame speed. Combust Flame 161(4):1125–30
4. Yelverton TLB, Roberts WL (2008) Effect of dilution, pressure, and velocity on smoke point in laminar jet flames. Combust Sci Technol 180(7):1334–1346
5. Acton MR, Baldwin PJ (2008) Ignition probability for high pressure gas transmission pipelines. In: Proceedings of the 7th international pipeline conference
6. Swuste P, van Nunen K, Reniers G et al (2019) Domino effects in chemical factories and clusters: an historical perspective and discussion. Process Saf Environ Prot 124:18–30

References

7. Scasta JD, Leverkus S, Tisseur D et al (2023) Vegetation response to a natural gas pipeline rupture fire in Canada's montane cordillera. Energy, Ecol Environ 8(5):457–470
8. Zhou K, Zhou M, Huang M et al (2022) An experimental study of jet fires in pits. Process Saf Environ Prot 163:131–143
9. Zhou K, Wang Y, Zhang L et al (2020) Effect of nozzle exit shape on the geometrical features of horizontal turbulent jet flame. Fuel 260:116356
10. Wang Z, Zhou K, Zhang L et al (2021) Flame extension area and temperature profile of horizontal jet fire impinging on a vertical plate. Process Saf Environ Prot 147:547–558
11. Wade RA, Sivathanu YR, Gore JP (1995) A study of two phase high liquid loading jet fires. National Institute of Standards and Technology, Gathersusbug, MD
12. Zhou K, Nie X, Wang C et al (2021) Jet fires involving releases of gas and solid particle. Process Saf Environ Prot 156:196–208
13. Shi X, Zhou K (2023) Geometrical features and global radiant heat of double turbulent jet flames. Fuel 350:128789
14. Hottel HC, Hawthorne WR (1948) Diffusion in laminar flame jets. Symp Combust & Flame & Explos Phenom 3(1):254–266
15. Hawthorne WR, Weddell DS, Hottel HC (1948) Mixing and combustion in turbulent gas jets. Symp Combust Flame, Explos Phenom 3(1):266–288
16. Becker HA, Liang D (1978) Visible length of vertical free turbulent diffusion flames. Combust Flame 32:115–137
17. Suris AL, Flankin EV, Shorin SN (1977) Length of free diffusion flames. Combust Explos Shock Waves 13(4):459–462
18. Sonju OK, Hustad J (1984) An experimental study of turbulent jet diffusion flames. Nor Marit Res 4(12):2–11
19. Palacios A, Muñoz M, Casal J (2009) Jet fires: an experimental study of the main geometrical features of the flame in subsonic and sonic regimes. AIChE J 55(1):256–263
20. Delichatsios MA (1993) Transition from momentum to buoyancy-controlled turbulent jet diffusion flames and flame height relationships. Combust Flame 92(4):349–364
21. Schefer RW, Houf WG, Williams TC et al (2007) Characterization of high-pressure, underexpanded hydrogen-jet flames. Int J Hydrog Energy 32(12):2081–2093
22. Bradley D, Gaskell PH, Gu X et al (2016) Jet flame heights, lift-off distances, and mean flame surface density for extensive ranges of fuels and flow rates. Combust Flame 164:400–409
23. Costa M, Parente C, Santos A (2004) Nitrogen oxides emissions from buoyancy and momentum controlled turbulent methane jet diffusion flames. Exp Therm Fluid Sci 28(7):729–734
24. Santos A, Costa M (2005) Reexamination of the scaling laws for NOx emissions from hydrocarbon turbulent jet diffusion flames. Combust Flame 142(1–2):160–169
25. Kiran DY, Mishra DP (2007) Experimental studies of flame stability and emission characteristics of simple LPG jet diffusion flame. Fuel 86(10–11):1545–1551
26. Gopalaswami N, Liu Y, Laboureur DM et al (2016) Experimental study on propane jet fire hazards: comparison of main geometrical features with empirical models. J Loss Prev Process Ind 41:365–375
27. Liu J, Fan Y, Zhou K et al (2018) Prediction of flame length of horizontal hydrogen jet fire during high-pressure leakage process. Procedia Eng 211:471–478
28. Zhou K, Liu J, Jiang J (2016) Prediction of radiant heat flux from horizontal propane jet fire. Appl Therm Eng 106:634–9
29. Vanquickenborne L, van Tiggelen A (1966) The stabilization mechanism of lifted diffusion flames. Combust Flame 10(1):59–69
30. Kalghatgi GT (1984) Lift-off heights and visible lengths of vertical turbulent jet diffusion flames in still air. Combust Sci Technol 41(1–2):17–29
31. Wu Y, Al-Rahbi IS, Lu Y et al (2007) The stability of turbulent hydrogen jet flames with carbon dioxide and propane addition. Fuel 86(12):1840–1848
32. Wu Y, Lu Y, Al-Rahbi IS et al (2009) Prediction of the liftoff, blowout and blowoff stability limits of pure hydrogen and hydrogen/hydrocarbon mixture jet flames. Int J Hydrog Energy 34(14):5940–5

33. Chamberlain GA (1987) Developments in design methods for predicting thermal radiation from flares. Chem Eng Res Des 65(4): 299–309
34. Markstein GH, de Ris J (1991) Wall-fire radiant emission. Part 1: Slot-burner flames, comparison with jet flames. Symp (Int) Combust 23(1):1685–1692
35. Turns SR, Myhr FH (1991) Oxides of nitrogen emissions from turbulent jet flames: Part I—Fuel effects and flame radiation. Combust Flame 87(3–4):319–335
36. Proust C, Jamois D, Studer E (2011) High pressure hydrogen fires. Int J Hydrog Energy 36(3):2367–2373
37. Molina A, Schefer RW, Houf WG (2007) Radiative fraction and optical thickness in large-scale hydrogen-jet fires. Proc Combust Inst 31(2):2565–2572
38. Schefer RW, Houf WG, Bourne B et al (2006) Spatial and radiative properties of an open-flame hydrogen plume. Int J Hydrog Energy 31(10):1332–1340
39. Zhou K, Qin X, Wang Z et al (2018) Generalization of the radiative fraction correlation for hydrogen and hydrocarbon jet fires in subsonic and chocked flow regimes. Int J Hydrog Energy 43(20):9870–9876
40. Fishburne ES, Pergament HS (1979) The dynamics and radiant intensity of large hydrogen flames. Symp (Int) Combust 17(1):1063–1073
41. Chamberlain GA (1987) Developments in design methods for predicting thermal-radiation from flares. Chem Eng Res Des 64(4):299–309
42. Brzustowski TA, Gollahalli SR, Gupta MP, et al (1975) Radiant heating from flares. In: Heat transfer conference, American society of mechanical engineers. ASME paper 75-HT-4, New York
43. McCaffrey B (1981) Some measurements of the radiative power output of diffusion flames. Western States Section Meeting, The Combustion Institute. Pittsburgh.: WSS/CU 81-15
44. Smith T, Periasamy C, Baird B et al (2005) Trajectory and characteristics of buoyancy and momentum dominated horizontal jet flames from circular and elliptic burners. J Energy Resour Technol 128(4):300–310
45. Zhang B, Liu Y, Laboureur D et al (2015) Experimental study on propane jet fire hazards: thermal radiation. Ind Eng Chem Res 54(37):9251–9256
46. Donaldson Cd, Snedeker RS (1971) A study of free jet impingement. Part 1. Mean properties of free and impinging jets. J Fluid Mech 45(2):281–319
47. Huang RF, Wang SM (1999) Characteristic flow modes of wake-stabilized jet flames in a transverse air stream. Combust Flame 117(1):59–77
48. Wang J-w, Fang J, Lin S-b, et al (2017) Tilt angle of turbulent jet diffusion flame in crossflow and a global correlation with momentum flux ratio. Proc Combust Inst 36(2):2979–2986
49. Gautam TK (1981) Blow-out stability of gaseous jet diffusion flames Part II: effect of cross wind. Combust Sci Technol 26(5–6):241–244
50. Gautam TK (1982) Blow-out stability of gaseous jet diffusion flames: Part III - effect of burner orientation to wind direction. Combust Sci Technol 28(5–6):241–245
51. Hu L, Zhang X, Zhang X et al (2014) A re-examination of entrainment constant and an explicit model for flame heights of rectangular jet fires. Combust Flame 161(11):3000–2
52. Zhou Z, Chen G, Zhou C et al (2019) Experimental study on determination of flame height and lift-off distance of rectangular source fuel jet fires. Appl Therm Eng 152:430–6
53. Iyogun CO, Birouk M (2008) Effect of fuel nozzle geometry on the stability of a turbulent jet methane flame. Combust Sci Technol 180(12):2186–209
54. Vijayan P, Sajeevan AC, Thampi GK et al (2024) Experimental evaluation of subsonic-horizontal jet flames: Impact of practical crack shapes. Fire Saf J 145:104127
55. Lattimer BY, Mealy C, Beitel J (2013) Heat fluxes and flame lengths from fires under ceilings. Fire Technology 49(2):269–291
56. Zhang X, Tao H, Zhang Z et al (2018) Flame extension area of unconfined thermal ceiling jets induced by rectangular-source jet fire impingement. Appl Therm Eng 132:801–807
57. Heskestad G, Hamada T (1993) Ceiling jets of strong fire plumes. Fire Saf J 21(1):69–82
58. Jessop DE, Jellinek AM (2014) Effects of particle mixtures and nozzle geometry on entrainment into volcanic jets. Geophys Res Lett 41(11):3858–3863

References

59. Apsley DD, Lane-Serff GF (2019) Collapse of particle-laden buoyant plumes. J Fluid Mech 865:904–27
60. Ramírez-Camacho JG, Pastor E, Amaya-Gómez R et al (2019) Analysis of crater formation in buried NG pipelines: a survey based on past accidents and evaluation of domino effect. J Loss Prev Process Ind 58:124–40
61. Zhou K, Wang Y, Simeoni A, et al (2024) Fire whirls induced by a line fire on a windward slope: a laboratory-scale study. Int J Wildland Fire 33(1):WF23048

Chapter 4
Radiant Heat Flux from Jet Diffusion Fire

Contents

4.1　Thermal Radiation Model of Jet Fires 68
　4.1.1　Point Source Radiation Model 68
　4.1.2　Multipoint Source Radiation Model 68
　4.1.3　Solid Flame Radiation Model 69
　4.1.4　Line Source Radiation Model 71
4.2　Comparisons of Radiant Heat Flux Between Model Predictions 74
　4.2.1　Radiant Heat Flux of Small Vertical Jet Fires 74
　4.2.2　Radiant Heat Flux of Medium Horizontal Jet Fires 77
　4.2.3　Radiant Heat Flux of Medium Vertical Jet Fires 79
　4.2.4　Radiant Heat Flux of Small Inclined Jet Fires 81
4.3　Application of Line Source Radiation Model 83
4.4　Summary 84
References 85

A jet fire following the rupture and leakage can produce a substantial amount of heat energy, posing a threat to the life and property in the vicinity. While its hazardous effects could confine to a relatively small area, it often triggers a series of events that can escalate the scale of the incident, namely, the domino effect. A survey performed on several accident data has revealed that a jet fire often acts as the initial stage in half of severe accidents such as explosions [1]. One primary mechanism deriving this severe process is the thermal radiation emitted by the jet flame, affecting both people and nearby facilities. Consequently, modelling thermal radiation from jet fires has been a significant focus in scientific and engineering research for decades and remains a major concern.

　　Current methods for modelling the thermal radiation of jet fires can be categorized into the field model and the semi-empirical model. The field model employs numerical technique to solve the differential equations governing mass, species, momentum and energy conservation in the flow and combustion process, also known as computational fluid dynamics (CFD). It offers a more rigorous and adaptable framework for addressing combustion and heat transfer problems compared to the semi-empirical

© The Author(s), under exclusive license to Springer Nature Singapore Pte Ltd. 2024　　67
K. Zhou, *Jet Fire Due to Gas Leakage*, Springer Series in Reliability Engineering,
https://doi.org/10.1007/978-981-97-5329-1_4

68   4 Radiant Heat Flux from Jet Diffusion Fire

model. However, the field model needs to incorporate empirical sub-models. On the other hand, the semi-empirical model facilitates rapid calculations for typical engineering issues. This chapter exclusively summarizes the engineering calculation methods for predicting the jet flame radiation.

There are four engineering type calculation methods, namely the point source radiation model, the multipoint radiation model, the solid flame radiation model and the line source radiation model [2]. These models are introduced in detail, followed by a discussion on how model predictions compare with measured radiant heat fluxes for jet fires of varying scales and directions. The chapter concludes with an application of the gas leakage model, jet flame combustion model, and line source radiation model to the transient hydrogen jet fire resulting from high-pressure leakage.

4.1 Thermal Radiation Model of Jet Fires

4.1.1 Point Source Radiation Model

All the flame radiation comes from a single point at the center of jet flame volume, which is the basic assumption of the point source radiation model. Such a point should be considered as a virtual origin with a radiant power equal to the product of the heat release rate and the radiative fraction, i.e. $\chi_r \dot{Q}$. Accordingly, the radiant heat flux ($\dot{q}''$) received by a nearby target can be calculated by

$$\dot{q}'' = \frac{\tau \chi_r \dot{Q}}{4\pi R_0^2} \cos\theta \qquad (4.1)$$

in which R_0 is the length of the connecting line between the target and the virtual origin, θ the angle between the connecting line and the normal direction of the target (as shown in Fig. 4.1), and τ the atmospheric transmissivity to the thermal radiation.

4.1.2 Multipoint Source Radiation Model

The multipoint source radiation model assumes that the thermal radiation stems from the multi points located in the centerline of the jet flame volume. Each point has the radiant power, and the radiant power sum of all points is $\chi_r \dot{Q}$. Therefore, the radiant heat flux from the multi point to a nearby target can be expressed by

$$\dot{q}'' = \sum_{j=1}^{N} \frac{w_j \tau_j \chi_r \dot{Q}}{4\pi R_j^2} \cos\theta_j \qquad (4.2)$$

4.1 Thermal Radiation Model of Jet Fires

Fig. 4.1 Schematic of the point source radiation model

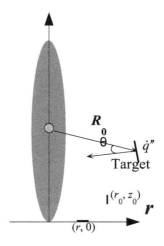

where the subscript j stands for the jth point source (as shown in Fig. 4.2), N is the total number of point sources and w_j is the weight of the jth point source. If $w_j = 1/N$, Eq. (4.2) can be named as the simple multipoint source radiation model, or otherwise it is a weighted multipoint source radiation model. Hankinson and Lowesmith [3] presented a correlation to calculate the w_j:

$$\left. \begin{array}{l} w_j = j w_1 \quad\quad\quad\quad\quad\quad\quad\quad\quad j = 1, \cdots, n \\ w_j = \left[n - \frac{n-1}{N-(n+1)} \cdot (j - (n+1)) \right] w_1 \quad j = n+1, \cdots, N \\ \sum_{1}^{N} w_j = 1 \end{array} \right\} \quad (4.3)$$

in which $n = 0.75N$. Obviously, the coefficient of 0.75 is empirical, and it could vary with the type and scale of jet fires.

4.1.3 Solid Flame Radiation Model

The solid flame radiation model considers that all the thermal energy is emitted from a simple yet realistical model of flame, in which a cone or a cylinder is often used to simulate the jet flame [4]. It predicts the radiant heat flux by

$$\dot{q}'' = \tau F E'' \quad (4.4)$$

In which F is the geometric view factor from flame to target, and E'' is the emissive power per flame surface area. The view factor depends on the assumed flame shape and the spatial relative position of the target against the flame. Some analytical correlations are available to calculate the view factor when the jet flame is simulated by a cylinder [5]. In addition, the contour integral method [6] or the direct numerical

Fig. 4.2 Schematic of the multipoint source radiation model

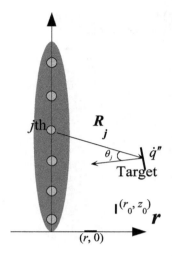

integration (e.g. the compound Simpson method [7]) can also calculate the view factor from a shape-complex flame to a target. The key issue is the increase of the computation cost as the jet flame is simulated using a complex geometric shape. A simple flame shape assumption could reduce the computation cost associated with the view factor, but it also overlooks some significant phenomena of jet flame (Fig. 4.3).

The emissive power of flame is proportional to the flame temperature (T_f) raised to the power of 4 multiplied by the flame emissivity (ε). The apparent flame emissivity increases to approach one as the flame thickness increases. In short, the emissive power of flame can be expressed by

$$E'' = \sigma T_f^4 (1 - \exp(-\kappa l_m)) \tag{4.5}$$

Fig. 4.3 Schematic of the solid flame radiation model

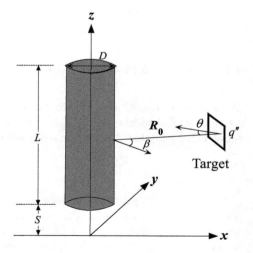

4.1 Thermal Radiation Model of Jet Fires

In which σ is Stefan–Boltzmann constant, i.e. 5.67×10^{-8} W/(m^2 K^4), κ is the absorption–emission coefficient of flame and l_m is the mean optical path of the entire flame volume. Another empirical method to calculate the emissive power of flame is

$$E'' = \chi_r \dot{Q} / A_f \qquad (4.6)$$

where A_f is the flame surface area.

In Eqs. (4.5) and (4.6), there is an implicit assumption that the emissive power is constant over the while flame surface. However, in the experimental investigation of vertical small jet fires, the emissive power first increases and then decreases as the height increases along the flame centerline [8]. Accordingly, three constant emissive powers are proposed for the continuous flame region, the intermittent flame region and the plume region, respectively [4], to approximate the variation of the emissive power over the flame surface.

4.1.4 Line Source Radiation Model

The line source radiation model assumes that all the radiant energy is emitted from the centerline inside the flame volume with the length equaling the flame length, as indicated in Fig. 4.4. The fact to support the assumption is the large ratio of the flame length to width for a fully turbulent jet fire. Therefore, the radiant heat flux from a jet flame to a nearby target can be quantified by

$$\dot{q}'' = \int_S^{S+L} \frac{\tau E'}{4\pi R^2} \cos \theta dz \text{ or } \dot{q}'' = \int_S^{S+L} \frac{\tau E'}{4\pi R^2} \cos \theta dx \qquad (4.7)$$

in which E' is the emissive power per line length (EPPLL). As shown in Fig. 4.4, the length R can be calculated by

$$R = \sqrt{(x - x_0)^2 + (y - y_0)^2 + (z - z_0)^2} \qquad (4.8)$$

The angle θ can be determined by

$$\cos \theta = \frac{(x - x_0)n_x + (y - y_0)n_y + (z - z_0)n_z}{R} \qquad (4.9)$$

For the vertical jet flame in Fig. 4.4a, $x = 0$ and $y = 0$, while $y = 0$ and $z = 0$ for the horizontal jet flame in Fig. 4.4b.

In general, the flame of hydrocarbon shows typical characteristics of volumetric emission and absorption due to high temperature gas and soot inside it. The flame is typically regarded as an isothermal, homogeneous grey emitter with a constant absorption–emission coefficient. Thus, Eq. (4.5) can quantify the emissive power per flame surface area. The mean optical path length in Eq. (4.5) is conventionally given by $l_m = 3.6 \, V/A$ in which V is the flame volume. For the infinitesimal volume indicated in Fig. 4.4, the mean optical path length can reduce to be $l_m = 1.8b$ in which

Fig. 4.4 Schematic of the line source radiation model

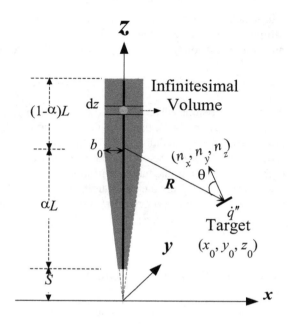

(a) Vertical jet flame

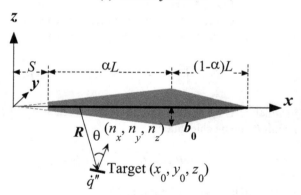

(b) Horizontal jet flame with high exit momentum

b is the radius or half width of flame. Therefore, Eq. (4.5) can derive the emissive power of the infinitesimal flame volume, as follows:

$$E'' = \sigma T_f^4 (1 - \exp(-1.8b\kappa)) \quad (4.10)$$

If the exponent $1.8b\kappa$ is small enough for the threadlike jet flame, Eq. (4.10) tells $E'' \sim b$. However, the assumption may fail in the case of a large jet fire involving heavily sooty fuel, as the flame radius and the absorption–emission coefficient could be sufficiently large.

4.1 Thermal Radiation Model of Jet Fires

In fact, the EPPLL stands for the emissive power of a certain point in the centerline, so the EPPLL should depend on the infinitesimal volume of flame with the corresponding point located inside, as indicated in Fig. 4.4. In mathematics, E' should equal $2\pi b E''$, which indicates that the EPPLL is proportional to the square of the flame radius, i.e. $E' \sim b^2$. Therefore, the EPPLL can be expressed by

$$E' = E_0' \cdot \left(b / b_o\right)^2 \tag{4.11}$$

where E_0' is the maximum EPPLL with the corresponding b_0 at the vertical distance of $S + \alpha L$ away from the leakage exit. Here, α is the ratio of the lower flame length to the whole flame length when the jet flame holds two different shapes respectively in the lower and upper portions (see Fig. 4.4). It is obvious that the length ratio α should be within 0–1.

Table 4.1 Geometric parameters of different jet flame shapes

Flame shape	b/b_0	$\int_S^{S+L} \left(b / b_0\right)^2 dz$
ellipse + ellipse	$\begin{cases} \sqrt{1 - \left(\frac{z-S-\alpha L}{S+\alpha L}\right)^2} & S \leq z \leq S + \alpha L \\ \sqrt{1 - \left(\frac{z-S-\alpha L}{L-\alpha L}\right)^2} & S + \alpha L \leq z \leq S + L \end{cases}$	$2L/3 + \alpha L/3 - (\alpha L)^3 \Big/ 3(S+\alpha L)^2$
cone + cone	$\begin{cases} \frac{z}{S+\alpha L} & S \leq z \leq S + \alpha L \\ \frac{S+L-z}{(1-\alpha)L} & S + \alpha L \leq z \leq S + L \end{cases}$	$(L+S)/3 - S^3 \Big/ 3(S+\alpha L)^2$
cylinder + cone	$\begin{cases} \frac{z}{S+\alpha L} & S \leq z \leq S + \alpha L \\ 1 & S + \alpha L \leq z \leq S + L \end{cases}$	$L + S/3 - \alpha L/3 - S^3 \Big/ 3(S+\alpha L)^2$

74   4 Radiant Heat Flux from Jet Diffusion Fire

Integrating the EPPLL over the full flame length leads to the total radiant heat output, namely, the product of the heat release rate and the radiative fraction. Thus, the maximum EPPLL can be calculated by

$$E_0' = \chi_r \dot{Q} \Big/ \int_S^{S+L} (b/b_0)^2 \mathrm{d}z \tag{4.12}$$

Obviously, Eqs. (4.11) and (4.12) can help to determine the EPPLL, if the jet flame shape is well known. Table 4.1 presents the geometric parameters in Eqs. (4.11) and (4.12) for three typically different jet flame shapes, namely, the back-to-back ellipse, the back-to-back cone and the combination of a cone in the lower portion and a cylinder in the upper portion. The three different shapes are considered to model the geometries of jet flames in typically different ranges of exit Froude number. The ellipse is suggested to model the shape of small-scale jet flames whose flame lengths varies from 28 mm to 52 mm [9]. The jet flame shape seems to evolve from an inverted cone attached to the burner rim to a cylinder detached from the burner rim, as the flame length is 440–592 mm [10]. Palacios and Casal [11] carried out experiments on large vertical propane sonic and subsonic jet fires, and proposed a cylinder to fit their large-scale jet flames of approximate 10 m in length. The combination of these different shapes could well approximate the jet flame.

4.2 Comparisons of Radiant Heat Flux Between Model Predictions

4.2.1 Radiant Heat Flux of Small Vertical Jet Fires

Baillie et al. [12] measured the horizontal and vertical profiles of the radiant heat flux of a methane jet flame whose exit diameter and velocity were 8.6 mm and 20 m/s, respectively. In detail, the horizontal heat flux sensors were flush with the jet exit, while the vertical heat flux sensors were 0.4 m away from the centerline of jet exit. Equation (3.3) predicts the flame Froude number of approximately 0.45, which indicates the flame dominated by buoyancy. Then Eq. (3.1) calculates the flame length of 1.14 m, and Eq. (3.5) gives the lift-off distance of 0.07 m. As indicated in Figs. 3. 2b and 3.3, its radiative fraction is determined to be 0.19. Different flame shapes are assumed to verify predictions, because no accurate shape has been reported for a vertical jet flame of approximately 1 m in length. Table 4.2 lists the parameters for all thermal radiation models to predict radiant heat flux.

The back-to-back ellipse flame shape is first used to facilitate the prediction of the line source radiation model. Figure 4.5 shows the model prediction versus the experimental measurement in terms of the horizontal and vertical profiles. The radiant heat fluxes are always overestimated near the leakage exit for the horizontal profile, and those at the middle positions are significantly underestimated for the vertical

4.2 Comparisons of Radiant Heat Flux Between Model Predictions

Table 4.2 Data of a small jet flame input into the line source model

d_2 (mm)	u_2 (m/s)	q_m (g/s)	$\dot{Q}$ (kW)	Fr	Fr_f	L (m)	S (m)	χ_r
8.6	20	0.75	32.05	69	0.45	1.14	0.07	0.19

[a] In calculation of q_m, 0.646 kg/m³ is used as the methane density; in calculation of $\dot{Q}$, 27.6 MJ/m³ is taken as the methane combustion heat and the combustion efficiency is of unity
[b] The density ratio of methane to ambient air is 0.55. The mass fraction of methane at stoichiometric condition is 0.0548. The methane adiabatic flame temperature rise is 1925 K at the ambient air temperature of 298 K

profile, despite the different length ratios. Thus, the back-to-back ellipse flame shape is not fit to simulate the small jet flame, in use of the line source radiation model.

Figure 4.6 depicts the prediction of the line source radiation model against the experimental measurement, when the combination of a cone and a cylinder models the flame shape. The model can give a good prediction as the length ratio is within 0.2–0.3 for the horizontal profile of radiant heat flux, but the prediction cannot keep consistency with the measured radiant heat flux in the vertical profile. Obviously, the cone-cylinder combined flame shape also cannot satisfy the line source radiation model, when it predicts thermal radiation from the small jet flame.

The prediction of the line source radiation model against the measurement are also given in Fig. 4.7, yet with the assumption of the back-to-back cone flame shape. The prediction of radiant heat flux firstly approaches to and then departures from the experimental measurement for both the horizontal and vertical profiles, as the length ratio increases. In particular, it can well agree with the experimental measurement, as the length ratio is approximately 0.5 indicating that the back-to-back cone is upper-lower symmetry. The nearly upper-lower symmetry flame shape of the back-to-back cone seems reasonable for the line source radiation model to predict the thermal radiation from the small jet fire.

Figure 4.8 presents the comparison of radiant heat flux between the experimental measurement and the predictions of different radiation models. In use of Eq. (4.1),

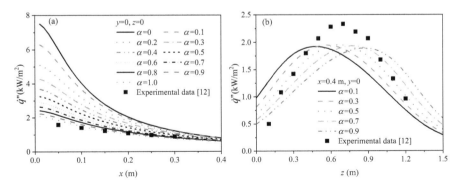

Fig. 4.5 Prediction of line source radiation model (ellipse + ellipse) against the experimental measurement: **a** horizontal and **b** vertical profiles of radiant heat flux [13]

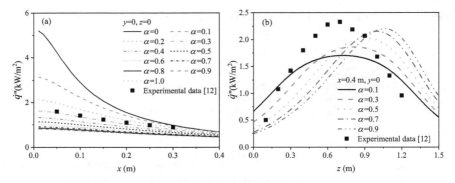

Fig. 4.6 Prediction of line source radiation model (cone + cylinder) against the experimental measurement: **a** horizontal and **b** vertical profiles of radiant heat flux [13]

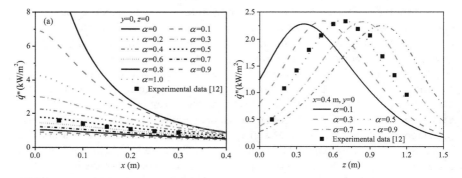

Fig. 4.7 Prediction of line source radiation model (cone + cone) against the experimental measurement: **a** horizontal and **b** vertical profiles of radiant heat flux [13]

the virtual origin of the point source radiation model is 0.64 m above the leakage exit. The 128-point sources are assumed to uniformly locate on the flame centerline, and Eq. (4.2) gives the prediction of the radiant heat flux, yet with the point source weight being 1/128 and calculated by Eq. (4.3), respectively, i.e., the simple and weighted multi-point source models. In use of the solid flame radiation model, the flame is assumed to be cylindrical, with its base positioned 0.07 m above the leakage exit and extending to a height of 1.14 m. The cylinder diameter is 0.16 m calculated by the correlation of $D/d = 3.55 Fr^{0.197}$ [14]. Thus, Eqs. (4.4) and (4.6) can be used to predict the radiant heat flux, with the geometric view factor correlations from a cylinder to the horizontal and vertical targets [5]. Note that the optimal ratio of 0.53 is selected for the back-to-back cone flame shape in use of the line source radiation model.

As shown in Fig. 4.8a, the radiant heat flux in the small horizontal positions, is significantly underestimated by the point source radiation model, while the radiant heat flux in the positions of over 0.4 m (more than the twice flame diameter) is well

4.2 Comparisons of Radiant Heat Flux Between Model Predictions

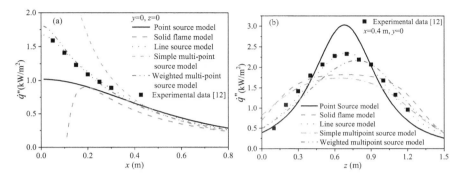

Fig. 4.8 Prediction of different radiation models against the experimental measurement: **a** horizontal and **b** vertical profiles of radiant heat flux [13]

predicted, which indicates its shortcomings for predictions in the near-field of jet flame. The solid flame radiation model cannot even give the right variation trend in the small horizontal positions, and gives an underestimation to some extent in the positions of over 0.2 m (a little more than the flame diameter). In contrast, the simple multipoint source radiation model significantly overestimates the prediction. The line source radiation model and the weighted multipoint source radiation model can agree well with the experimental measurement.

As shown in Fig. 4.8b, the point source radiation model significantly overestimates the radiant heat fluxes in the middle positions near the flame center, for these positions are very near the virtual origin, and underestimates those in the other positions. In contrast, both the solid flame radiation model and the simple multipoint source radiation model underestimate and overestimate the radiant heat flux in the middle positions and the other positions, respectively. This indicates that the emissive power per flame surface/length cannot be assumed to be constant across the entire jet flame surface/length. The weighted multipoint source radiation model gives a considerable prediction, but the line source radiation model can better fit the experimental measurement.

The big difference in prediction between different models could be explained, by the fact that the combustion energy distributes in the entire flame volume. The underlying mechanism of a good model is to simulate the distribution of thermal radiant energy to a reasonable and considerable extent. That is to say, the line source radiation model can better quantify the radiant energy distribution of jet flame than the other four models.

4.2.2 Radiant Heat Flux of Medium Horizontal Jet Fires

Zhang et al. [15] recently conducted an experimental measurement on the radiant heat flux from the horizontal propane jet fires of 2–5 m in length. Seven radiometers

78   4 Radiant Heat Flux from Jet Diffusion Fire

(R1–2, R5–6 and R9–11) are positioned with the surface normal orientation of n_x = −0.174, n_y = −0.985 and n_z = 0, if the gas jet direction coincides with x-axis, as described in Fig. 4.4b. Table 4.3 shows the coordinates of radiometers and their time-averaged readings. The radiant heat flux decreases from 5.62 kW/m^2 to 0.62 kW/m^2 as the y_0 increases from 1.83 m (near the flame) to 6.53 m (far away from the flame). Note that 5 kW/m^2 is suggested to be the critical radiant heat flux for distinguishing the near and far fields. The data of d_2 = 19.1 mm and u_2= 206 m/s is used to validate the model, for its large exit momentum results in the horizontal centerline of the jet flame.

The propane mass leakage rate of 0.11 kg/s and its combustion heat of 46 kJ/g can give the heat release rate of 5078 kW, assuming the combustion efficiency of unity. Equation (3.1) calculates the flame length of 4.95 m, and Eq. (3.5) gives the lift-off distance of 1.97 m. The radiative fraction is estimated to be 0.11 by the laws in Fig. 3.2c or Fig. 3.3. These four calculated parameters, together with the radiometer position and surface normal orientation, are input into the thermal radiation models for predicting the radiant heat flux against the experimental measurement.

The virtual origin of the point source radiation model is at the center of flame, i.e. (3.96, 0, 0) in the Cartesian coordinates. The 128-point sources uniformly locate along the x-axis from 1.97 m to 4.95 m for the multipoint source radiation model. The weight of each point is considered to be 1/128 or calculated by Eq. (4.3). The back-to-back cone of 0.4–0.6 in length ratio is used to simulate the horizontal jet flame shape, in use of the line source radiation model.

Figure 4.9 shows the comparison of the radiant heat fluxes between model predictions and experimental measurements. In the far field, all the models predict radiant heat fluxes without big difference. In the near field, the point source radiation model and two multipoint source radiation models overestimate the radiant heat flux, while the line source radiation model agrees with the experimental measurement. In short, the line source radiation model with a reasonable flame shape assumption can also give a good description of the thermal radiant energy distribution of horizontal jet fires.

Table 4.3 Radiant heat flux of medium horizontal jet fires [16]

Radiometer	x_0 (m)	y_0 (m)	z_0 (m)	$\dot{q}''_{experiment}$ (kW/m^2)
R1	5.54	3.19	−0.4	2.70
R2	2.81	4.99	−0.4	1.80
R5	4.08	2.43	−0.4	5.19
R6	1.44	3.00	−0.4	2.81
R9	5.20	1.83	−0.4	5.62
R10	2.67	6.53	−0.4	0.62
R11	4.45	4.50	−0.4	2.16

[a]The test condition: d_2 = 19.1 mm, u_2 = 206 m/s and q_m = 0.11 kg/s

4.2 Comparisons of Radiant Heat Flux Between Model Predictions

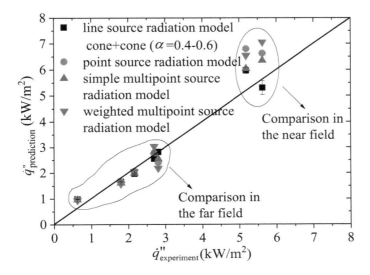

Fig. 4.9 Prediction versus measurement of radiant heat flux from medium horizontal jet fires: comparison between different radiation models [16]

4.2.3 Radiant Heat Flux of Medium Vertical Jet Fires

Palacios et al. [4] conducted a serial of large outdoor experiments on propane jet flames of different orifice exit diameters and mass flow rates in still air. The radiant

Table 4.4 Parameters of medium vertical jet flames for the line source radiation model

d_2 (mm)	q_m (g/s)	u_2 (m/s)	$\dot{Q}$(kW)	Fr	Fr_f	L (m)	S (m)	χ_r	$\dot{q}''$(kW/m^2) at (x_0, y_0, z_0)		
									(1, 0, 0.5)	(3, 0, 1)	(5, 0, 1)
43.1	500	184	23,000	282	1.58	9.70	0.63	0.14	10.60	7.02	4.96
	470	173	21,620	265	1.49	9.55	0.59	0.15	10.39	7.14	5.06
30	340	258	15,640	396	2.14	7.30	0.99	0.09	9.11	5.08	3.05
	310	235	14,260	361	1.96	6.89	0.90	0.10	8.95	4.85	2.80
	300	227	13,800	350	1.90	6.50	0.88	0.10	8.90	4.85	2.73
	250	189	11,500	291	1.61	6.48	0.70	0.13	8.83	4.73	2.64

[a]In calculation of u_2, 1.868 kg/m^3 is used as the propane density; in calculation of $\dot{Q}$, 46 kJ/g is used as the propane combustion heat and the combustion efficiency is of unity; the unit of r_0 and z_0 is meter

[b]The density ratio of propane to ambient air is 1.52. The mass fraction of propane at stoichiometric condition is 0.0599. The propane adiabatic flame temperature rise is 1955 K at the ambient air temperature of 298 K

Fig. 4.10 Prediction versus measurement of radiant heat flux from medium vertical jet fires [13]

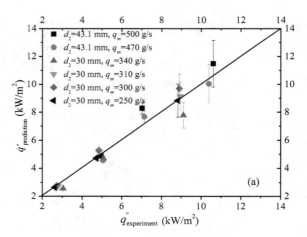

(a) line source radiation model

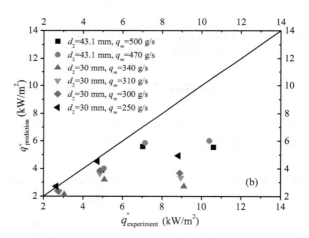

(b) weighted multipoint source radiation model

heat fluxes were measured at three positions during all tests. They also presented the infrared images of jet flame and thereby suggested a cylinder approximating the flame shape. However, the flame image showed a fast increase of flame diameter with respect to the vertical distance in the lower portion, indicating the flame shape resembling a type of the cone-cylinder combined shape with a little length ratio. The radiative fractions can be estimated by the laws in Fig. 3.2c or Fig. 3.3. The flame length and lift-off distance are estimated by Eqs. (3.1) and (3.5), respectively. Table 4.4 lists the input parameters for the line source radiation model to predict the radiant heat flux from the large jet flames. As shown, the flame Froude number is much larger than that of the above small jet flame.

4.2 Comparisons of Radiant Heat Flux Between Model Predictions

In calculation, the length ratio α is assumed to be from 0.1 to 0.2 with step of 0.02 due to its uncertainty, and then a group of radiant heat flux data can be obtained by Eq. (4.7) with the other input parameters listed in Table 4.4. Figure 4.10a shows the radiant heat flux comparison between the prediction of line source radiation model and the experimental measurement. The bars in Fig. 4.10a indicate the standard deviation of the prediction due to the uncertainty of the length ratio. The cone-cylinder combined flame shape seems reasonable for the line source radiation model to well predict the radiant heat flux of medium jet flames, even though the exponent of $1.8b\kappa$ could be large due to the fact that the medium jet flames hold a considerable flame radius. The weighted multipoint source radiation model ($N = 1280$) is also used to calculate the radiant heat flux from medium jet flames. As shown in Fig. 4.10b, it gives a significant underestimation in the near field, even though it can well predict the radiant heat flux from the small vertical jet flame (see Sect. 4.2.1). In brief, the line source radiation model can also accurately predict the radiant heat flux from medium jet flames.

In addition to the length ratio, the other input parameters could also have the uncertainty that results from the obscure definition and calculation method, as well as the measurement uncertainty, especially for the large and transient jet fire test. By parameter sensitivity and uncertainty analysis, the ranking by importance of input parameters is the flame length, the heat release rate, the radiative fraction, the length ratio, and the lift-off distance for the line source radiation model [7]. The ranking helps to know the priority that should be given to the accuracy of the leakage and flame behavior correlations.

4.2.4 Radiant Heat Flux of Small Inclined Jet Fires

The jet flame would tilt away from the jet direction to complicate the flame shape, when the flame buoyancy and the exit momentum are not coincident in the direction. The complex flame shape would complicate the calculation of the virtual point position in the point source radiation model, the weight in the multipoint source radiation model, and the view factor in the solid flame radiation model. The flame shape depicted in Fig. 4.4b is reasonable only when the exit momentum is much over the flame buoyancy. Accordingly, there are two categories of line source radiation models for the radiant heat flux from inclined jet fires [17]. The first one only takes into account the effect of exit momentum, and assumes the straight centerline for the jet flame, while the other considers both the flame buoyancy and the exit momentum, and assumes the curved centerline for the jet flame. Figure 4.11 depicts the radiant heat transfer from an inclined jet flame to the ruptured tank surface. As shown, the flame centerline gradually departs away from the jet direction during the development, and two-line segments of BE and BC are used to approximate the curved centerline. If let BE and BC equal the flame length and zero, respectively, the flame centerline coincides with the jet direction.

82   4 Radiant Heat Flux from Jet Diffusion Fire

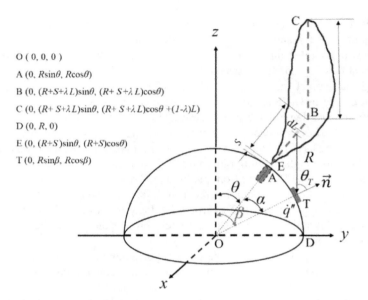

Fig. 4.11 Schematic of radiant heat transfer from an inclined jet flame to a target on the surface of the ruptured tank [17]

The difference between the two radiation models is whether the flame buoyancy is considered or not. As the exit Froude number increases, the role of flame buoyancy decreases as compared to that of exit momentum. The comparisons between the predictions of the two different radiation models against the experimental measurement are presented in Fig. 4.12, for three different exit Froude numbers. For $Fr = 29$, the model I underestimates the measurement, while the model II well agrees with the measurement, as shown in Fig. 4.12a. No doubt is that the flame buoyancy dominates the flame behavior as $Fr = 29$. As the exit Froude number increases to be 9000, the exit momentum could increase to be comparable to the flame buoyancy. However, the model II still gives better predictions than the model I, as indicated by Fig. 4.12b. As the exit Froude number continues to increase, the role of exit momentum would be significantly greater than that of flame buoyancy. Accordingly, the model II reduces to be the model I, as evidenced by the little deference of predictions between two models in Fig. 4.12c.

4.3 Application of Line Source Radiation Model

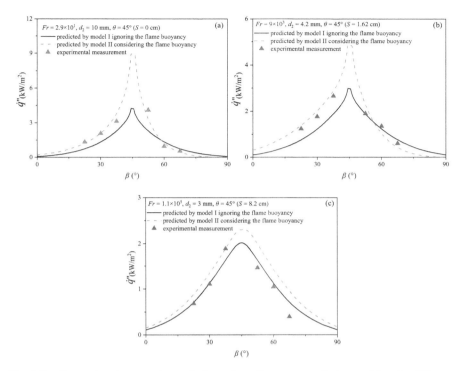

Fig. 4.12 Comparison between the predictions of two line source radiation model under different exit Froude numbers [17]

4.3 Application of Line Source Radiation Model

The radiant heat fluxes of large hydrogen jet fires measured by Proust et al. [18], are used to test the line source radiation model. The flame length calculated by the Fr_f number correlation (Eqs. (3.1)–(3.3)) as shown in Fig. 3.6d, and the lift-off distance calculated by Eq. (3.5) with the flow velocity and diameter in Fig. 3.6a, are input into the line source radiation model (Eqs. (4.7)–(4.12)). The input parameters also need the heat release rate and the radiative fraction in Fig. 3.8a. The radiative fractions calculated by the correlations based on the global residence time (Eq. (3.10)) and the Fr_f number (Eq. (3.11)), respectively, are input into the radiation model for comparison in terms of the predicted radiant heat flux. In use of the line source radiation model, the flame shape of horizontal hydrogen jet fires is modeled by a pair of back-to-back cones with a length ratio of 0.8.

Figure 4.13 shows the comparison between model prediction and experimental measurement of the radiant heat flux versus the leakage time. The line source radiation model underestimates all the radiant heat fluxes at the five measurement points, when the radiative fraction calculated by Eq. (3.10) is used, as shown in Fig. 4.13a. However, the model prediction well agrees with the experimental measurement at

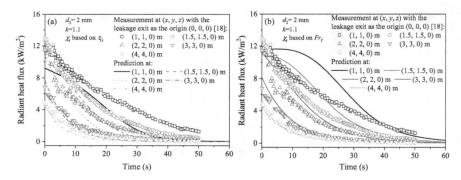

Fig. 4.13 Comparison in the variation of radiant heat fluxes with the leakage time between experimental measurements and model predictions: radiative fraction correlation based on the **a** global residence time (Eq. (3.10) and **b** Fr_f number (Eq. (3.11))

the positions of (3, 3, 0) m and (4, 4, 0) m, while it overestimates the radiant heat flux at the other positions, for the use of the radiative fraction determined by Eq. (3.11), as indicated in Fig. 4.13b. The overestimation could result from the variation of jet flame shape with the leakage time, in contrast to the assumption of the constant flame shape. Anyway, Eq. (3.11) based on Fr_f number could give a more reliable prediction on the radiative fraction than Eq. (3.10) based on the global residence time, in terms of the prediction accuracy and safety design of the radiant heat flux.

4.4 Summary

The radiant heat flux of jet fires is of great interest for evaluating the risk or damage to nearby people and property. Four different engineering types of calculation methods are reviewed in detail, and their predictions are compared to the measured radiant heat flux from jet fires of different scales and jet directions. Finally, the line source radiation model is suggested to predict the transient field of radiant heat flux from a high-pressure hydrogen jet fire. The key points are:

(1) In use of the line source radiation model, the back-to-back ellipse, the back-to-back cone, and the cone-cylinder combined shape are proposed to approximate the jet flame geometry, as the scale of the jet fire increases. This shape approximation assists the model in extensively considering the nonuniform distribution of the radiant heat energy within the flame volume.
(2) The line source radiation model provides a better prediction of the radiant heat flux, as compared to the point source radiation model, the multipoint source radiation model, and the solid flame radiation model.

(3) The correlation based on Fr_f number (Eq. (3.11)) seems to better predict the radiative fraction than that based on the global residence time (Eq. (3.10)), from the viewpoint that the radiative fraction, as an input parameter of the line source radiation model, aids in calculating the radiant heat flux.

References

1. Gómez-Mares M, Zárate L, Casal J (2008) Jet fires and the domino effect. Fire Saf J 43(8):583–8
2. Zhou K, Jiang J (2016) Thermal radiation from vertical turbulent jet flame: line source model. J Heat Transf 138(4):042701
3. Hankinson G, Lowesmith BJ (2012) A consideration of methods of determining the radiative characteristics of jet fires. Combust Flame 159(3):1165–77
4. Palacios A, Muñoz M, Darbra RM et al (2012) Thermal radiation from vertical jet fires. Fire Saf J 51:93–101
5. Mudan KS (1984) Thermal radiation hazards from hydrocarbon pool fires. Prog Energy Combust Sci 10(1):59–80
6. Sparrow EM (1963) A new and simpler formulation for radiative angle factors. J Heat Transf 85(2):81–7
7. Zhou K, Liu N, Zhang L et al (2014) Thermal radiation from fire whirls: revised solid flame model. Fire Technology 50(6):1573–87
8. Markstein GH, De Ris J (1991) Wall-fire radiant emission. Part 1: Slot-burner flames, comparison with jet flames. Symp (Int) Combust 23(1):1685–1692
9. Zhang X, Hu L, Zhu W et al (2014) Flame extension length and temperature profile in thermal impinging flow of buoyant round jet upon a horizontal plate. Appl Therm Eng 73(1):15–22
10. Kiran DY, Mishra DP (2007) Experimental studies of flame stability and emission characteristics of simple LPG jet diffusion flame. Fuel 86(10–11):1545–51
11. Palacios A, Casal J (2011) Assessment of the shape of vertical jet fires. Fuel 90(2):824–33
12. Baillie S, Caulfield M, Cook DK et al (1998) A Phenomenological model for predicting the thermal loading to a cylindrical vessel impacted by high pressure natural gas jet fires. Process Saf Environ Prot 76(1):3–13
13. Apsley DD, Lane-Serff GF (2019) Collapse of particle-laden buoyant plumes. J Fluid Mech 865:904–27
14. Bagster DF, Schubach SA (1996) The prediction of jet-fire dimensions. J Loss Prev Process Ind 9(3):241–5
15. Zhang B, Laboureur DM, Liu Y et al (2018) Experimental study of a liquefied natural gas pool fire on land in the field. Ind Eng Chem Res 57(42):14297–306
16. Zhou K, Liu J, Jiang J (2016) Prediction of radiant heat flux from horizontal propane jet fire. Appl Therm Eng 106:634–9
17. Wu Y, Zhou K, Zhou M et al (2022) Radiant heat feedback from a jet flame to the ruptured tank surface. Int J Therm Sci 172:107322
18. Proust C, Jamois D, Studer E (2011) High pressure hydrogen fires. Int J Hydrog Energy 36(3):2367–73

Chapter 5
Theoretical Framework for Calculating Jet Fire Risk Induced by High-Pressure Transient Leakage

Contents

5.1 Theoretical Framework of Gas Leakage, Jet Flame, Thermal Radiation and Damage
  Criteria 89
  5.1.1 Damage Criteria for People and Structures: Radiant Heat Flux Threshold 91
  5.1.2 Damage Criteria for People and Structures: Thermal Dose 93
  5.1.3 Damage Criteria for People and Structures: Target Temperature 94
5.2 Parameter Sensitivity and Uncertainty Analysis of Input Parameters 95
  5.2.1 Parameter Sensitivity Analysis of the Theoretical Framework 96
  5.2.2 Uncertainty Analysis on the Results of the Theoretical Framework 97
5.3 Validation of Theoretical Framework 100
  5.3.1 Case 1: Large Hydrogen Jet Fire 100
  5.3.2 Case 2: Large Natural Gas Jet Fire 103
  5.3.3 Case 3: Large Hydrogen/Natural Gas Mixture Jet Fire 107
5.4 Summary 108
References 110

The high-pressure technique reduces the costs associated with the gas storage and transportation. Consequently, the jet fires that result from high-pressure gas leaks have garnered significant attention from scientists and engineers. Figure 5.1 illustrates the schematic of the high-pressure leakage process, the subsequent jet flame combustion, the heat transfer process, the thermal response of target, and the key parameters that describe all these processes.

Experimental studies have been conducted intensively on full-scale jet fires due to the high-pressure gas transient leakage. Schefer et al. [2] conducted experiments on the time profiles of the pressure and temperature inside the tank, with an initial release pressure up to 413 bar, and measured the flame length and radiant heat flux of the hydrogen jet fire. Imamura et al. [3] measured the hydrogen mass flow rate, flame length and width, temperature field of jet fires as the release pressure decreased. Studer et al. [4] measured the flame length and radiant heat flux of jet fires due to the transient leakage of the hydrogen and hydrogen/methane mixture, with initial pressures of 100 bar. The mass flow rate, flame length and radiant heat flux field versus

© The Author(s), under exclusive license to Springer Nature Singapore Pte Ltd. 2024
K. Zhou, *Jet Fire Due to Gas Leakage*, Springer Series in Reliability Engineering,
https://doi.org/10.1007/978-981-97-5329-1_5

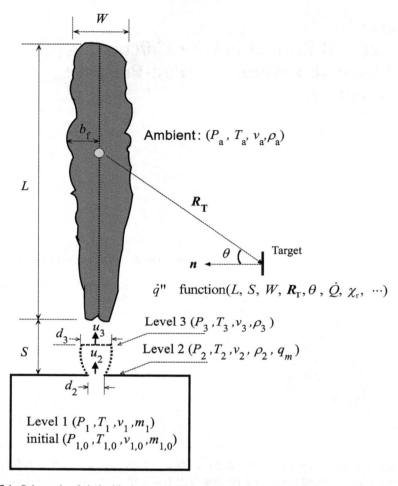

Fig. 5.1 Schematic of choked leakage and subsequent jet flame [1]

the leakage time were measured for the large hydrogen jet fire with an initial pressure of 61 bar [5]. The pressure and temperature inside the tank and at the exit, the mass flow rate, jet flame length and radiant heat flux were experimentally investigated for hydrogen jet fires, as the tank pressure dropped from the initial 900 bar [6]. Lowesmith and Hankinson [7] measured the mass flow rate, flame length and radiant heat flux field of large natural gas and hydrogen/natural gas mixture jet fires with an initial pressure of over 70 bar. However, these experimental tests are highly case-specific, and the general laws cannot be extrapolated from these results. Moreover, only a subset of parameters in Fig. 5.1 were measured, likely due to the economic costs and operation challenges associated with large-scale testing.

5.1 Theoretical Framework of Gas Leakage, Jet Flame, Thermal Radiation … 89

Numerical simulations have also been conducted to study the flame geometry and radiation characteristics of high-pressure jet fires [8–10]. However, these simulations typically assume a constant mass leakage rate. Clearly, the current numerical simulation methods do not consider the coupling process of leakage and combustion, and they only predict the key parameters of the flame combustion and heat transfer process (Fig. 5.1). In addition, these simulations generally consist of many sub-models, and the choice of different sub-models can significantly affect the accuracy of the numerical simulation [11].

This chapter presents a theoretical framework for predicting the behaviors and properties of large jet fires resulting from high-pressure gas leakage. The theoretical framework consists of the gas leakage model and the notional nozzle model from Chap. 2, the jet flame combustion model from Chap. 3, the thermal radiation model from Chap. 4, and the thermal damage criteria for people and structures, as summarized from available literature. The robustness of the theoretical framework is validated by field test results of the hydrogen, natural gas, and hydrogen/natural gas mixture jet fires. This chapter also provides a ranking of input parameters by importance for the theoretical framework through parameter sensitivity analysis, finally followed by a probit analysis of jet fire accident consequences using uncertainty analysis.

5.1 Theoretical Framework of Gas Leakage, Jet Flame, Thermal Radiation and Damage Criteria

Figure 5.2 shows the logic and academic idea of the theoretical framework. If the initial conditions inside the storage tank and the leakage exit size are known, the theoretical framework can predict the state properties, flow parameters, flame size and radiant heat flux field versus the leakage time, and the thermal damage on a target. It consists of a transient leakage model, a notional nozzle model, a flame length correlation, a lift-off distance correlation, a radiative fraction correlation, a thermal radiation model, and damage criterion on people and structure in terms of radiant heat flux threshold, thermal dose and target temperature.

The model based on the van der Waals equation of state in Sect. 2.2.3, is suggested to quantify the transient leakage. The notional nozzle model is introduced in detail in Sect. 2.3. The correlation based on Fr_f number, i.e. Equations (3.1)–(3.3) in Sect. 3.1.1, helps to calculate the flame length of jet fire, while Eq. (3.5) in Sect. 3.1.2 can determine the lift-off distance. The correlation based on Fr_f number, i.e. Equation (3.11) and Fig. 3.3 in Sect. 3.1.3, is used to calculate the radiative fraction. The line source radiation model in Sect. 4.1.4, is suggested to predict the thermal radiation from large jet fires.

In addition, the physicochemical property of the leaked gas should be determined before the use of the theoretical framework. The physical parameters of typical pure gases are listed in Table 2.1, and Eqs. (2.46)–(2.50) in Sect. 2.4 help to extrapolate

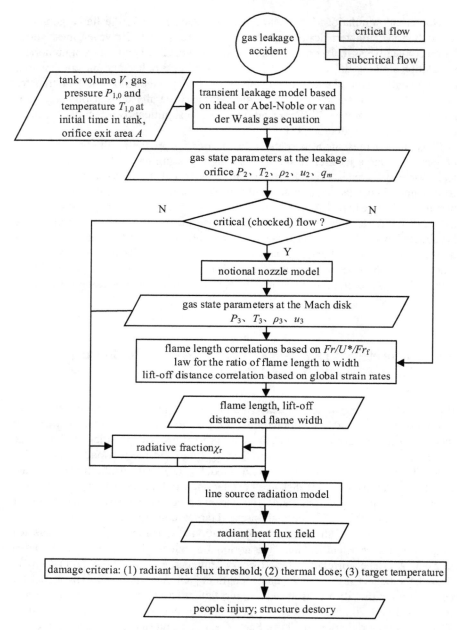

Fig. 5.2 Theoretical framework for calculating the high-pressure leakage and subsequent jet fire. It is updated from Zhou et al. [1]

5.1 Theoretical Framework of Gas Leakage, Jet Flame, Thermal Radiation ... 91

Table 5.1 Chemical parameters of common gas

Fuel	ΔH (kJ/kg)	$T_{ad}{}^a$ (K)	f_s
H_2	142×10^3	2382	0.0283
CH_4	50×10^3	2336	0.0550
C_2H_6	52×10^3	2330	0.0588
C_3H_8	46×10^3	2327	0.0602

[a]The adiabatic flame temperature is calculated using the software of HPFLAME

the physical property of mixture gas from those of pure gases. The chemical property of leaked gas involves the heat of combustion, the adiabatic flame temperature and the mass fraction of fuel at stoichiometric conditions. Table 5.1 lists the chemical parameters of typical pure gases. The heat of combustion can be calculated for the mixture gas by

$$\Delta H = \sum_{i=1}^{\eta} w_i \Delta H_i \qquad (5.1)$$

The mass fraction of mixture fuel at stoichiometric conditions can be determined by

$$f_s = \sum_{i=1}^{\eta} w_i \Delta f_{si} \qquad (5.2)$$

The hydrogen jet fire of Acton et al. [5], and the natural gas and hydrogen/natural gas mixture jet fires of Lowesmith and Hankinson [7] are used to validate the theoretical framework. The natural gas consisted of 93% CH_4, 5% C_2H_6, 0.3% C_3H_8 and 1.7% N_2 by volume, while the hydrogen/natural mixture consisted of 22.3% hydrogen and 77.7% natural gas by volume [7]. With the physical parameters in Table 2.1 and the chemical parameters in Table 5.1, Eqs. (2.46)–(2.50) and (5.1), (5.2) can calculate the physicochemical parameters of natural gas and hydrogen/natural gas mixture, as listed in Table 5.2.

5.1.1 Damage Criteria for People and Structures: Radiant Heat Flux Threshold

The damage of thermal radiation on people and structures depends on the radiant heat flux and the duration of exposure. The vulnerability of humans is also influenced by the ease of sheltering, the individual differences, and the type of clothing worn. The structure component includes the wood, plastic, steel, brick, etc. The thermal

Table 5.2 Physicochemical parameters input into the theoretical framework

Fuel	R_g (J/ (kg·K))	a ((m^6·Pa)/ kg^2)	b (m^3/ kg)	k	ΔH (kJ/ kg)	$T_{ad}{}^a$ (K)	$\Delta T_{ad}/T_a{}^c$	f_s	ρ/ρ_a
H$_2$	4157	6084	0.0132	1.40	1.42×10^5	2382	7.0	0.0283	0.069
NG	489	843	0.0026	1.30	4.87×10^4	2334	7.3	0.0569	0.589
H$_2$/ NG	609	1032	0.0031	1.32	5.18×10^4	2337	7.4	0.0551	0.471

[a]The adiabatic flame temperature is calculated using the software of HPFLAME, and the ambient temperature is assumed to be 298 K

radiation exposure could ignite the flammable materials and reduce the mechanical integrity of structure. The vulnerability of humans and structure to the thermal radiation can be calculated by different threshold values of radiant heat flux with diverse consequences. Table 5.3 summarizes thermal radiation exposure effects over a range of radiant heat fluxes. Notice that the information provided in Table 5.3 can be expected in typical situations alongside guidance on how other factors could enhance or reduce the effects.

The risk-based approach using the threshold requires the constant radiant heat flux. However, the high-pressure transient leakage leads to a jet fire whose combustion intensity decreases with the leakage time, and thus the decrease of the radian heat flux with time. Apparently, the transient characteristic challenges the application of the threshold in Table 5.3 into the risk analysis of the high-pressure jet fire.

Table 5.3 Thermal radiation exposure effects [12, 13]

Radiant heat flux (kW/m^2)	Effects
1.4	Harmless for persons without any special protection
1.7	Minimum required to cause pain
2.1	Minimum required to cause pain after 60s
4.0	Causes pain after an exposure of 20s (first degree burns)
7.0	Maximum tolerable value for firefighters completely covered protected by special Nomex protective clothes
10	Certain polymers can ignite
11.7	Thin steel (partly insulated) can lose mechanical integrity
12.6	Wood can ignite after a long exposure
16	Blistering of exposed skin after 5s
20	Piloted wood ignition after more than 5.5 min
25	Thin steel (insulated) can lose mechanical integrity
37.5	Damage to process equipment and collapse of mechanical structures

5.1 Theoretical Framework of Gas Leakage, Jet Flame, Thermal Radiation …

5.1.2 Damage Criteria for People and Structures: Thermal Dose

The risk-based approach using the thermal dose mathematically integrates the radiant heat flux and the exposure duration. The thermal dose (D) to quantify the people injury can be expressed by

$$D = \dot{q}''^{4/3}t \tag{5.3}$$

As the radiant heat flux varies with the time, Eq. (5.3) can be written in the integral form as

$$D = \int_0^t \dot{q}''^{4/3}\mathrm{d}t \tag{5.4}$$

The pain, first, second and third degree burn thresholds of thermal dose are in the range of 86–103 $(\mathrm{kW/m^2})^{4/3}$s, 80–130 $(\mathrm{kW/m^2})^{4/3}$s, 240–350 $(\mathrm{kW/m^2})^{4/3}$s, and 870–2600 $(\mathrm{kW/m^2})^{4/3}$s, respectively, which is determined by experimental burn test of infrared radiation [14]. Someone could argue that the pain occurs at approximately 80–100 $(\mathrm{kW/m^2})^{4/3}$s, the second degree burns at 240–730 $(\mathrm{kW/m^2})^{4/3}$s, and the third degree burns at approximately 1000 $(\mathrm{kW/m^2})^{4/3}$s [15]. The thermal dose range and its difference between different literatures show the uncertainty of test due to many potential factors. Therefore, a probit approach is proposed to further quantify the risk of thermal dose. In the probit approach, the probability of fatality takes the form

$$P = \int_{-\infty}^{Y-5} \frac{1}{\sqrt{2\pi}} e^{\frac{-u^2}{2}} \mathrm{d}u, \; Y = a + b\ln(D) \tag{5.5}$$

Different values of a and b are suggested in the available literature. Eisenberg et al. [16] gave $a = -4.9$ and $b = 2.56$ derived from the data of the Hiroshima nuclear explosion incident whose thermal heat radiation is in the ultraviolet spectrum. Lately, Tsao and Perry [17] accounted for the difference between ultraviolet and infrared radiation in hydrocarbon fires, and kept $b = 2.56$ but modified $a = -2.8$ by increasing the thermal dose by a factor of 2.23. Equation (5.5) can also describe the possibility of the first, second, third degree burns, but with different values for parameters a and b.

The effect of time-varying radiant heat on structure ignition can be quantified by the integration of the radiant heat flux with time. Cohen and Butler [13] proposed the concept of the flux-time integral (FTP). FTP can be derived from the time profile of radiant heat flux by

$$\text{FTP} = \int_{t_0}^{t_f} \delta(\dot{q}'' - \dot{q}''_{cr})^{1.828} dt \tag{5.6}$$

where $\dot{q}''_{cr}$ is the minimum incident heat flux to cause ignition, t_0 and t_f are the initial and final heat exposure times, respectively, and the coefficient δ equals 0 for $\dot{q}'' \leq \dot{q}''_{cr}$ or 1 for $\dot{q}'' > \dot{q}''_{cr}$. The ignition occurs as the FTP increases to reach the minimum FTP for ignition (FTP$_{ig}$). $\dot{q}''_{cr} = 13.1$ kW/m^2 and FTP$_{ig} = 11,501$ kJ/m^2 for wood [13]. The variation of $\dot{q}''_{cr}$ and FTP$_{ig}$ with material type limits the use of Eq. (5.6).

5.1.3 Damage Criteria for People and Structures: Target Temperature

In essence, the people injury and structure destroy result from the increase of temperature under the exposure of radiant heat. The skin burns, i.e., the pain and damage to the skin, occurs when the temperature is over 44 °C at the basal layer. Wieczorek and Dembsey [18] have summarized four algorithms that can model the skin temperature over time. It should be stressed that the skin is assumed to be a single layer, opaque simi-infinite solid in each of these models. With the known temperature–time (T–t) history of the skin, the Arrhenius formular can help to calculate the injury parameter (Ω) by

$$\Omega = \int_0^t A_e \exp(-\Delta E / RT) dt \tag{5.7}$$

where A_e is pre-exponential term, ΔE is activation energy, and R is the universal gas constant. Different values of Ω stands for different levels of skin burns. For example, $\Omega = 0.53$ represents the first-degree burn, while $\Omega = 1$ is the superficial second-degree burn.

The simi-infinite solid could simulate some structure components that hold a large surface area and the conductive heat transfer only in the direction of thickness. Under the effect of radiant heat, the net heat flux on the surface of structure component is assumed to approach the radiant heat flux from the jet fire, without the consideration of the convection and reradiation cooling effect. Then the surface temperature of structure component (T_s) can be expressed by

$$T_s = T_{s0} + \frac{2\dot{q}''(\alpha t / \pi)^{0.5}}{k_c} \tag{5.8}$$

in which T_{s0} is the surface temperature at the initial time, α and k_c are the thermal diffusivity and conductivity of the structure component. Equation (5.8) requires the constant radiant heat flux. For the time-varying radiant heat, Eq. (5.8) can be written in the integral form as

$$T_s = T_{s0} + \frac{(\alpha/\pi)^{0.5}}{k_c} \int_0^t \dot{q}'' t^{-0.5} dt \tag{5.9}$$

The radiant heat from high-pressure jet fires could ignite the tree leaves and grasses near the structure and then flame spread towards the structure. The leaf and grass hold a small volume due to little thickness, and thus the conductive heat transfer inside them can be neglected. The lumped capacitance method can help to calculate the temperature of the little leaf and grass by

$$T_{st} = T_{st0} + \frac{A_{st}}{\rho_{st} V_{st} c_{st}} \int_0^t \dot{q}'' dt \tag{5.10}$$

where T_{st}, A_{st}, ρ_{st}, V_{st} and c_{st} are the temperature, area of surface heated by radiation, density, volume and heat capacity of the small target. The ratio of V_{st} to A_{st} is the thickness of the small target.

5.2 Parameter Sensitivity and Uncertainty Analysis of Input Parameters

The input parameters could have considerable error or uncertainty in full-scale tests or accidents, probably resulting from the obscure definition and calculation method, as well as the enhancement of measurement uncertainty with the increase of scale. Accordingly, two key points should be clear as the theoretical framework is used to quantify the fire risk of high-pressure gas leakage. One is to clarify the ranking by importance of all input parameters, and the other is to quantify the uncertainty of jet fire accidental consequence due to the uncertainty of mainly important input parameters. The local sensitivity analysis method and the Monte Carlo simulation are introduced to explore the two key points, respectively.

The application of the theoretical framework into the natural gas pipeline leakage is taken as the example. Table 5.4 lists the working pressure and diameter of some natural gas transmission pipelines in China. The spacing between two gas transmission stations is often 24 km in GB 50,251–2015 *Code for Design of Gas Transmission Pipeline Engineering*. The median value is selected as reference for the sensitivity analysis. In detail, the broken pipeline holds the initially absolute pressure of 8.25 MPa, the diameter of 0.83 m and the length of 12 km.

Table 5.4 Working pressure and diameter of natural gas pipelines in China

Name of pipeline	Working absolute pressure (MPa)	Diameter (mm)
west-to-east gas pipeline I	10.1	1016
west-to-east gas pipeline П	10.1	1219
west-to-east gas pipeline Ш	10.1	1219/1016
Sichuan to east gas pipeline	10.1	1016
Shanxi to Beijing gas pipeline	6.5/10.1	660/1016
Synthetic natural gas pipeline in Fuxin	6.4	450/600/900
Sino-Middle Asia gas pipeline	6.4	610
Sino-Myanmar gas pipeline	10.1	1016

5.2.1 Parameter Sensitivity Analysis of the Theoretical Framework

The local sensitivity analysis can give the relative uncertainty (S_ps) of thermal dose versus the input parameter, expressed by

$$S_p(D, x_i) = \frac{\partial D}{\partial x_i} \Delta x_i / D \times 100\% \qquad (5.11)$$

The input parameter (x_i) includes the working pressure, diameter and length of the gas transmission pipeline, the ambient wind speed and pressure. Two typical wind speeds of 5 m/s and 10 m/s and one standard atmosphere pressure are regarded as reference values for sensitivity analysis.

The transient leakage model based on van der Walls equation, the flame length and radiative fraction correlations based on Fr_f, and the line source radiation model are selected to construct the theoretical framework. The theoretical framework can calculate the radiant heat flux versus time at different target positions away from the broken pipeline under the conditions of the reference working pressure, pipeline diameter and length, ambient wind speed and pressure. Under the wind effect, the whole jet flame is assumed to incline with the tilt angle calculated by Eq. (3.12), and then the line source radiation model is used to calculate the radiant heat flux of inclined jet flame, as discussed in Sect. 4.2.4.

The time-profiles of radiant heat flux are input into Eq. (5.4) to calculate the thermal dose versus the target position. In calculation of the thermal dose causing people injury by Eq. (5.4), the exposure time of 30s is considered. The consideration is the significant decrease of radiant heat flux after 30s leakage. In addition, the people could evacuate away from the leakage point.

The reference position of target that corresponds to the thermal dose causing the 50% lethality is calculated by Eq. (5.5). At the reference position, the relative variation of the thermal dose is determined by Eq. (5.11), by varying one input parameter away from the reference value by 20% and keeping the other parameters

5.2 Parameter Sensitivity and Uncertainty Analysis of Input Parameters

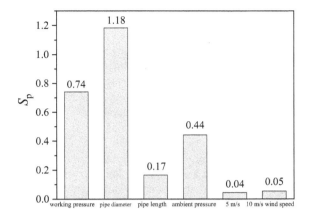

Fig. 5.3 Relative uncertainty of the thermal dose versus the input parameter. The target position where the thermal dose causes 50% lethality is considered for sensitivity analysis

constant. Figure 5.3 shows the relative uncertainty of the thermal dose at the reference position resulting from uncertainties of the five input parameters. The ranking by importance of the five input parameters is the pipeline diameter, the working pressure, the ambient pressure, the pipeline length, and the ambient wind speed. The wind speed holds little effect on the uncertainty of thermal dose, for it is comparatively little against the high-pressure leakage velocity within the initial 30s, and thus the tilt angle is little according to Eq. (3.12). In short, the length and working pressure of pipeline are considered to affect the uncertainty of thermal dose in the Monte Carlo simulation.

5.2.2 Uncertainty Analysis on the Results of the Theoretical Framework

Monte Carlo simulation method uses random numbers to perform calculations in a computer. Given the probability density functions of the input parameters, it samples the input parameters and then computes the result for each sample. A sufficient number of samples would lead to the frequency distribution of results, which can help to explore the characteristics of the probability distribution.

The specific steps of Monte Carlo simulation are as follows: (1) Determine the uncertain parameters input into the model and their probability distribution characteristics. (2) Generate a certain number of input samples based on the distribution characteristics of input parameters. (3) Input the generated samples into the model for computation. (4) Analyze the calculation results to understand the probability distribution characteristics of the output results.

The sampling method can affect the result of the Monte Carlo simulation. The random sampling (RS) and the Latin Hypercube sampling (LHS) are two commonly used sampling methods. The LHS divides the cumulative distribution of input parameters into equal intervals, and then randomly samples within each interval. Obliviously, the LHS would produce the better sample whose structure is similar to that of population than the RS, if the sample number is same.

As listed in Table 5.4, the working absolute pressure of natural gas pipeline ranges from 6.4 MPa to 10.1 MPa, and the pipe diameter is from 0.45 m to 1.22 m. The worst scenario is considered that the pipeline is fully ruptured. The absolute gas pressure and the ruptured pipe diameter are assumed to fellow the uniform distribution in their respective ranges. Figure 5.4 displays the frequency distribution of samples for the absolute gas pressure under different sample sizes using RS and LHS. Comparison between Fig. 5.4a and b, shows that the sample size of RS significantly affects the distribution, and that the distribution approaches to be uniform as the sample size increases. However, the distribution of sample is uniform under a relatively small sampling size, as evidenced by the comparison between Fig. 5.4c and d.

The increase of sampling size increases the accuracy of calculation result, but it also increases the cost of calculation source. Accordingly, the balance between the calculation accuracy and cost is of great importance. The target position where the thermal dose causes 50% lethality is also considered for uncertainty analysis. Figure 5.5 shows the mean and variance of the distance between the target position and the leakage point. As the sampling size increases to be over 10,000, the LHS can

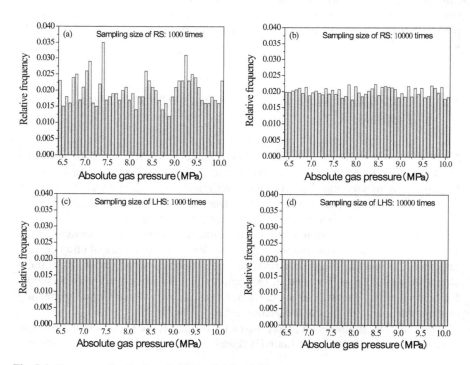

Fig. 5.4 Frequency distribution of samples for the absolute gas pressure

5.2 Parameter Sensitivity and Uncertainty Analysis of Input Parameters   99

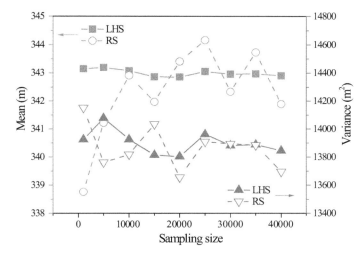

Fig. 5.5 Mean and variance of the target position where the thermal dose causes 50% lethality away from the leakage point, as a funtion of the sampling size

quickly give a relatively stable mean and variance, while those of RS significantly varies. In short, the LHS and the sampling size of 10,000 would be used for the following uncertainty analysis.

The 10,000 samples of working absolute pressures and pipeline diameters, obtained by LHS method, are input into the theoretical framework. The target positions where the thermal doses cause 1%, 50% and 99% lethality are as the calculation results of the theoretical framework. Table 5.5 shows the distance between the target position and the leakage point. The maximum working pressure and diameter of natural gas pipeline, i.e., 10 MPa and 1.29 m, give the maximum distance, while the 6.3 MPa and 0.45 m result in the minimum distance. However, the mean distance is not the average of the maximum and minimum.

Figure 5.6 shows the probability density of the distance away from the leakage point to cause the 1%, 50% and 99% lethality. The cumulative probability is also presented. The mean and variance of the distance away from the leakage point are used to build the normal density function and the lognormal density function. As shown, the probability density significantly increases, and then almost keeps constant, and finally decreases, as the distance increases. The lognormal density function can

Table 5.5 Distances between target positions of different fatalities and leakage point

Lethality (%)	Maximum (m)	Minimum (m)	Mean (m)
1	918	209	519.5
50	630	122	343.1
99	519	80	260

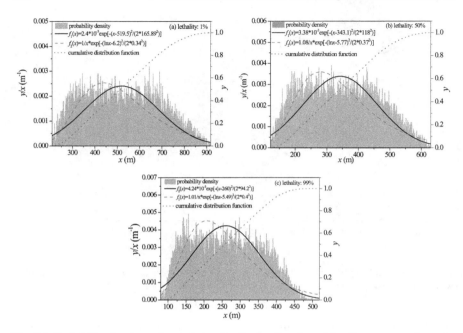

Fig. 5.6 Probability density and cumulative distribution function of the distance away from the leakage point to cause **a** 1%, **b** 50% and **c** 99% lethality

better follow the increasing trend than the normal density function, but the final decreasing trend can be better modeled by the normal density function.

5.3 Validation of Theoretical Framework

5.3.1 Case 1: Large Hydrogen Jet Fire

Acton et al. [5] conducted full-scale experiments of hydrogen jet fires due to the release of the tank of 61 bar in initial pressure and 163 m^3 in volume. The high-pressure tank was buried in soil backfill. A pipeline of 15.24 cm in diameter was connected to the tank. The pipeline was ruptured by the detonation, resulting in a full-bore leak where the exit diameter matched the pipeline diameter. The gas mass flow rate was indirectly calculated by the measurements on the static pressure at the pipe wall, the stagnation pressure on the centerline of the pipe and the gas temperature near the release point. The flame geometry was recorded by two infrared thermal imaging cameras. The radiant heat flux of jet fire was measured by four wide-angle radiometers that were 40 m, 50 m, 60 m, and 80 m away from the flame centerline, respectively. The position and orientation of radiometers are redrawn in Fig. 5.7. The

5.3 Validation of Theoretical Framework

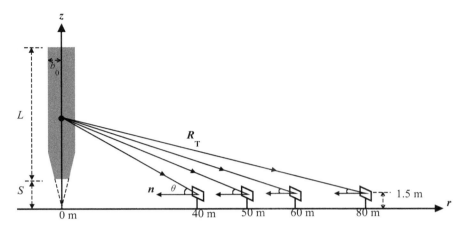

Fig. 5.7 Schematic of the radiometer positions for the hydrogen jet fire [1]

test repeated twice, and the one in the relatively light wind was used to validate the theoretical framework.

The time-averaged radiative fraction of the hydrogen jet fire was measured to be 0.29 during the whole leakage, for the sand and soil entrained into the flame volume played a positive effect. The initial condition parameters input into the gas transient leakage model are listed in Table 5.6. In calculation, the ambient pressure and temperature are assumed to be 1.01 bar and 298 K, respectively.

The gas transient leakage model in Sect. 2.2.3 is used to calculate the mass flow rate of hydrogen, with the physicochemical parameters in Table 5.2 and the initial condition parameters in Table 5.6. Figure 5.8 shows the model prediction of the mass flow rate versus the leakage time against the experimental measurement. The model firstly overestimates the gas mass flow rate, and then gradually approaches to the measurement as the leakage time goes. However, the prediction gives a shorter duration of the whole leakage than the measurement. The significant overestimation within the 0–10 s could be explained by the appearance of "necking" on the ruptured pipeline end. Due to the necking phenomenon, the gas leakage area was less than the pipe cross-sectional area of the full-bore release. That is to say, the actual leakage area was less than that of 15.24 cm in diameter, which leads to the overestimation of the mass flow rate in the initial period. Anyway, the relative error is estimated to be 20.5 ± 13.6% for $d_2 = 15.24$ cm. However, as the assumed leakage area decreases, the leakage time increases and the prediction gradually approaches to the measurement.

Table 5.6 Initial condition parameters for the hydrogen transient leakage

$P_{1,0}$ (MPa)	$T_{1,0}$ (K)	V (m^3)	d_2 (cm)	$v_{1,0}$ (m^3/kg)	$m_{1,0}$ (kg)	$v_{cr,0}$
6.1	298	163	15.24	0.210	775.45	0.503

Fig. 5.8 Comparison of the hydrogen mass flow rate versus the leakage time between model prediction and experimental measurement [1]

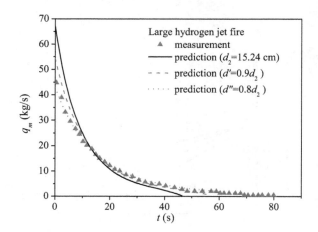

In particular, the prediction well agrees with the measurement for the leakage exit of $0.8d_2$ in diameter.

The flame length versus the leakage time is predicted by Eqs. (3.1)–(3.3) with the leakage model output parameters and the physicochemical parameters of hydrogen in Table 5.2. Figure 5.9 presents the prediction of the total flame height versus leakage time against the measurement. It should be stressed that the current model can predict the decay of flame length with the leakage time. However, the growth of a jet flame length is the fact in the initial period. Accordingly, the comparison between prediction and measurement is done after the flame growth time. As shown, this model first well fellows the measurement, and then gives an underestimation, as the leakage time goes over 20 s. This could result from the fact that the gas mass flow rate predicted by the leakage model decreases more rapidly than the measurement when the leakage time is over 20 s. That's to say, the uncertainty of the leakage model prediction significantly affects the prediction of jet flame length correlation in the later release period. The relative error is estimated to be 17.7 ± 14.8% for d_2 = 15.24 cm. It is also found that the prediction well agrees with the measurement as the leakage exit is of $0.8d_2$ in diameter. Note that Eqs. (3.1)–(3.3) only give the mean flame length and cannot fellow the actual pulsation of flame length. In addition, Eqs. (3.1)–(3.3) are limited to the jet fires in still air and with the circular leakage exit.

The radiant heat flux of jet fire can be predicted by the line source radiation model into which are input by the results of gas leakage and jet flame size models. Figure 5.10 shows the model prediction of radiant heat flux against the experimental measurement. In detail, the prediction well agrees with the measurement in the positions of over 50 m, but overestimates the radiant heat flux at the near positions of less than 50 m. This may result from the assumption of the time-averaged radiative fraction for model calculation. The relative errors are estimated to be 24.1 ± 27.0% and 14.7 ± 14.4% for the time of 12.5 s and 20.5 s, respectively. Moreover, the

5.3 Validation of Theoretical Framework

Fig. 5.9 Comparison of the hydrogen flame length with the leakage time between model prediction and experimental measurement [1]

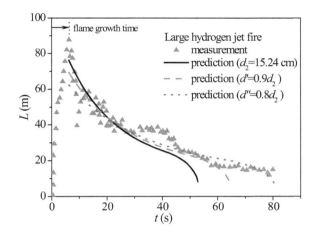

Fig. 5.10 Comparison of the radiant heat flux between model prediction and experimental measurement [1]

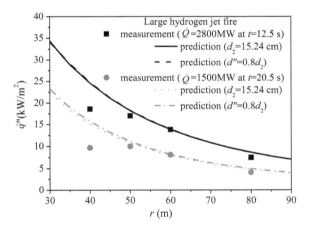

uncertainty of the leakage exit area seems to little affect the prediction of radiant heat flux.

5.3.2 Case 2: Large Natural Gas Jet Fire

Lowesmith and Hankinson [7] recently conducted two large-scale experiments to measure the jet fire following the rupture of high-pressure pipeline conveying natural gas and hydrogen/natural gas mixtures, respectively. The pipeline was connected to a storage tank buried in a crater. The tank was 163 m^3 in volume, and the pipeline was 15 cm in diameter. To initiate the leakage and jet fire, an explosive charge was used to cause complete rupture of the pipeline. The initially absolute pressure and temperature in the storage tank were 71.5 bar and 281 K, respectively. The ambient wind speed was approximately 5–6 m/s in test. Table 5.7 lists the initial condition parameters for the model to predict the natural gas jet fire. The total gas mass flow rate was indirectly calculated by the measured pressure and temperature at an orifice plate installed in the pipeline leading from tank. The jet flame length was determined using two cameras positioned for horizontal viewing, one in the south and the other in the east. The radiant heat flux distribution was measured using Medtherm radiometers with wide-angle lenses. In order to ensure that all the flame could be seen by the radiometer, the normal direction of each radiometer surface was adjusted to face the point approximately 60 m above the leakage exit. Figure 5.11 shows the position and orientation angle of each radiometer. The ambient pressure and temperature are assumed to be 1.01 bar and 298 K, respectively.

The mass flow rate of the natural gas can be precited by the van der Waals leakage model in Sect. 2.2.3, with the physicochemical parameters in Table 5.2 and the initial condition parameters in Table 5.7. Figure 5.12 shows the comparison between prediction and measurement. The prediction overestimates the mass flow rate at the initial time, and then gradually approaches the measurement as the leakage time goes. The relative error is estimated to be 40.6 ± 20.4% for $d_2 = 15$ cm. The significant deviation between prediction and measurement could result from the fact that the actual leakage exit size is less than 15 cm in diameter. As shown, the prediction gradually approaches to the measurement as the leakage exit reduces to be of $0.7d_2$ in diameter. The determination method of the mass flow rate could be inaccurate

Table 5.7 Initial condition parameters for the natural gas transient leakage model

$P_{1,0}$ (MPa)	$T_{1,0}$ (K)	V (m^3)	d_2 (cm)	$v_{1,0}$ (m^3/kg)	$m_{1,0}$ (kg)	$v_{cr,0}$
7.15	281	163	15	0.016	10,516	0.477

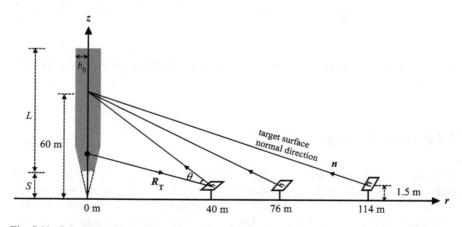

Fig. 5.11 Schematic of the radiometer positions for the natural gas or hydrogen/natural gas mixture jet fires [1]

5.3 Validation of Theoretical Framework

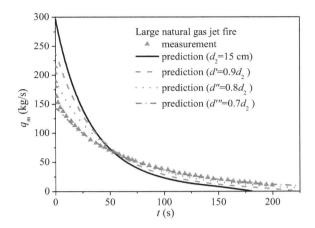

Fig. 5.12 Comparison of the gas mass flow rate versus leakage time between model prediction and experimental measurement (natural gas) [1]

as stressed in [7], for it was deduced by the pipe stagnation pressure and the gas temperature at the position located a certain distance from the leakage exit. That may be why the relatively big deviation exists in the initial period.

The flame length can be predicted by the correlation based on Fr_f (Eqs. (3.1)–(3.3)), with the output parameters of the leakage model and the natural gas physicochemical parameters in Table 5.2. Figure 5.13 shows the prediction of the flame length against the measurement. The prediction first significantly overestimates the flame length and then gradually approaches to the measurement. The overestimation could result from the uncertainty of leakage model prediction in the initial leakage period. Note a short period for the flame growth in the beginning. Obviously, the model fails to give predictions in the flame growth period. In addition, the jet flame considerably fluctuates with time due to the buoyancy effect, which can also enlarge the relative error between prediction and measurement. The relative errors are estimated to be 22.5 ± 17.0% and 29.8 ± 25.7%, as the predictions are compared to the measurements of south and east cameras, respectively, as $d_2 = 15$ cm. However, the prediction would approach to the measured flame length as the leakage exit size decreases.

The radiant heat flux field of jet fire can be calculated by the line source radiation model, with the calculation results of the leakage and flame size models and also the radiative fraction. In particular, the variation of radiative fraction with the leakage time can be predicted by Eq. (3.11) and Fig. 3.3 in Sect. 3.1.3, with the natural gas physicochemical parameters in Table 5.2, as shown in Fig. 5.14. Figure 5.15 shows the prediction of the radiant heat flux versus leakage time against the measurement. No obvious differences in the prediction of radiant heat flux are found at different distances, and the prediction considerably agrees with the measured radiant heat flux. That is to say, the theoretical framework can give an estimation on the variation of thermal radiation field to some extent during the high-pressure natural gas leakage process. The relative errors are estimated to be 36.9 ± 22.4%, 39.0 ± 19.9% and 39.6 ± 19.2% for the radial distances of 40 m, 76 m and 144 m, respectively.

Fig. 5.13 Comparison of the flame length versus leakage time between model prediction and experimental measurement (natural gas) [1]

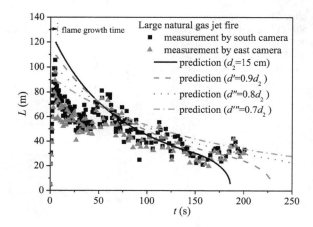

Fig. 5.14 Radiative fraction versus leakage time (natural gas) [1]

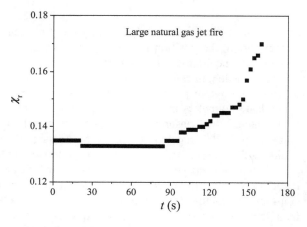

Fig. 5.15 Comparison of the radiant heat flux versus leakage time between model prediction and experimental measurement at different distances away from the natural gas jet fire [1]

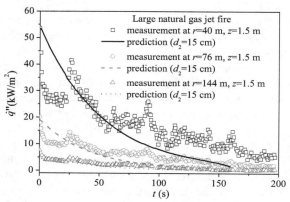

5.3 Validation of Theoretical Framework

5.3.3 Case 3: Large Hydrogen/Natural Gas Mixture Jet Fire

The hydrogen/natural gas mixture jet fire test in [7] is also used for further validation of the theoretical framework. At the initial time, the absolute pressure and temperature in the storage tank were 72.6 bar and 277 K, respectively. The ambient wind speed was estimated to be 1–2 m/s in test. The other initial condition parameters, as well as the measurement method on mass flow rate, flame height and radiant heat flux, are the same as those of natural gas jet fire in Sect. 5.3.2.

The gas mass flow rate versus the leakage time is calculated by the van der Waals leakage model in Sect. 2.2.3. The model calculation requires the physicochemical parameters in Table 5.2, and the initial condition parameters in Table 5.8. Figure 5.16 shows the comparison of the hydrogen/natural gas mass flow rate between the prediction and measurement. The leakage model firstly gives an overestimation for the gas mass flow rate, and then the prediction seems to keep good consistency with the measurement. Obviously, the deviation between prediction and measurement for the large hydrogen/natural gas mixture jet fire is similar to that of the natural gas jet fire, for the same measurement method on mass flow rate was used. The relative error is estimated to be $35.7 \pm 19.7\%$ for $d_2 = 15$ cm. There is no doubt that the model prediction would also approach to the experimental measurement as the leakage exit decreases to be of $0.7d_2$ in diameter.

Figure 5.17 depicts the flame length versus leakage time between model prediction and experimental measurement. The prediction can keep good consistency with the measurement during the leakage process. The relative errors are estimated to be 15.6

Table 5.8 Initial condition parameters for the mixture gas transient leakage model

$P_{1,0}$ (MPa)	$T_{1,0}$ (K)	V (m^3)	d_2 (cm)	$v_{1,0}$ (m^3/kg)	$m_{1,0}$ (kg)	$v_{cr,0}$
7.26	277	163	15	0.020	8150	0.481

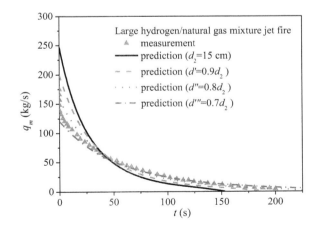

Fig. 5.16 Comparison of the gas mass flow rate versus leakage time between model prediction and experimental measurement (hydrogen/natural gas mixture) [1]

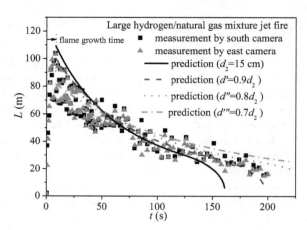

Fig. 5.17 Comparison of the flame length versus leakage time between model prediction and experimental measurement (hydrogen/natural gas mixture) [1]

± 12.9% and 12.2 ± 11.5% for the departure of the prediction from the measurements of south and east cameras, respectively. The prediction also further approaches to the measured flame length as the leakage exit area decreases. In addition, the theoretical model seems to give a better prediction on the flame length of the mixture jet fire than that of the nature gas jet fire. Note the less ambient wind for the hydrogen/natural gas mixture jet fire. Thus, it would be of great interest to take into account the ambient wind for the theoretical framework.

The radiative fraction of natural gas/hydrogen mixture jet fires can also be determined by the radiative fraction correlation, i.e., Eq. (3.11) and Fig. 3.3 in Sect. 3.1.3, with the physicochemical parameters of hydrogen/natural gas mixtures in Table 5.2. As shown in Fig. 5.18, the radiative fraction firstly decreases to a little extend and then significantly increases. The line source radiation model is used to calculate the variation of radiant heat flux field with the leakage time. Figure 5.19 presents the comparison of radiant heat flux between prediction and measurement at different distances away from the jet fire. In the near field, the prediction can well agree with the measurement. However, the line source radiation model underestimates the radiant heat flux in the far field. The relative errors are estimated to be 30.7 ± 17.8%, 32.6 ± 23.1% and 40.2 ± 23.3% for the radial distances of 40 m, 76 m and 144 m, respectively.

5.4 Summary

A theoretical framework is proposed to calculate the damage on people and structure caused by high-pressure jet fires. This framework consists of the gas leakage model, the notional nozzle model, the jet flame length correlation, the radiative fraction correlation, the line source radiation model, and the thermal damage criteria for people and structures. The parameter sensitivity analysis is used to rank the input

5.4 Summary

Fig. 5.18 Radiative fraction versus leakage time (hydrogen/natural gas mixture) [1]

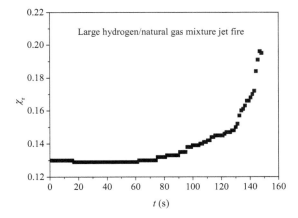

Fig. 5.19 Comparison of the radiant heat flux versus leakage time between model prediction and experimental measurement at different distances away from the hydrogen/natural gas mixture jet fire [1]

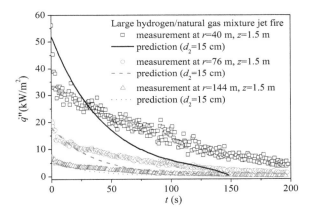

parameters by importance. The ranking order is as follows: the diameter of ruptured pipeline (leakage exit area), the operating pressure of pipeline, the ambient pressure, the pipeline length, and the ambient wind speed.

The Monte Carlo simulation method is used to quantify the uncertainty in the consequences calculated by the theoretical framework. Over 10,000 samples are required for the Latin Hypercube sampling method to ensure accuracy. Apart from the thermal damage criteria, the robustness of the other sub-models within the theoretical framework is validated by field test results of the hydrogen, natural gas, and hydrogen/natural gas mixture jet fires. In Chap. 6, the theoretical framework will be applied to the consequence analysis of jet fire accidents resulting from the rupture of high-pressure natural gas pipelines.

References

1. Zhou K, Wang X, Liu M et al (2018) A theoretical framework for calculating full-scale jet fires induced by high-pressure hydrogen/natural gas transient leakage. Int J Hydrogen Energy 43(50):22765–22775
2. Schefer RW, Houf WG, Williams TC et al (2007) Characterization of high-pressure, underexpanded hydrogen-jet flames. Int J Hydrogen Energy 32(12):2081–2093
3. Imamura T, Hamada S, Mogi T et al (2008) Experimental investigation on the thermal properties of hydrogen jet flame and hot currents in the downstream region. Int J Hydrogen Energy 33(13):3426–3435
4. Studer E, Jamois D, Jallais S et al (2009) Properties of large-scale methane/hydrogen jet fires. Int J Hydrogen Energy 34(23):9611–9619
5. Acton M R, Allason D, Creitz LW et al (2010) Large scale experiments to study hydrogen pipeline fires. In: Proceedings of the 8th international pipeline conference, Calgary, Alberta, Canada
6. Proust C, Jamois D, Studer E (2011) High pressure hydrogen fires. Int J Hydrogen Energy 36(3):2367–2373
7. Lowesmith BJ, Hankinson G (2013) Large scale experiments to study fires following the rupture of high pressure pipelines conveying natural gas and natural gas/hydrogen mixtures. Process Saf Environ Prot 91(1–2):101–111
8. Brennan SL, Makarov DV, Molkov V (2009) LES of high pressure hydrogen jet fire. J Loss Prev Process Ind 22(3):353–359
9. Wang CJ, Wen JX, Chen ZB et al (2014) Predicting radiative characteristics of hydrogen and hydrogen/methane jet fires using FireFOAM. Int J Hydrogen Energy 39(35):20560–20569
10. Consalvi J-L, Nmira F (2019) Modeling of large-scale under-expanded hydrogen jet fires. Proc Combust Inst 37(3):3943–3950
11. Papanikolaou E, Baraldi D, Kuznetsov M et al (2012) Evaluation of notional nozzle approaches for CFD simulations of free-shear under-expanded hydrogen jets. Int J Hydrogen Energy 37(23):18563–18574
12. Zárate L, Arnaldos J, Casal J (2008) Establishing safety distances for wildland fires. Fire Saf J 43(8):565–575
13. Cohen JD, Butler BW (1996) Modeling potential structure ignitions from flame radiation exposure with implications for wildland/urban interface fire management. In: Proceedings of the thirteenth fire and forest meteorology conference, Lorne, Australia
14. O'Sullivan S, Jagger S (2004) Human vulnerability to thermal radiation offshore. Harpur Hill, Buxton, Derbyshire: Health and Safety Laboratory
15. Purser DA (1997) Review of human response to thermal radiation, HSE contract research report No. 97/1996: S. M. Hockey and P. J. Rew. HSE Books, PO Box 1999, Sudbury, Suffolk CO10 6FS, UK 1996, 49 pp., £15·00 net, ISBN 0 7176 1083 7, paperback. Fire Safety J 28(3): 290–291
16. Eisenberg NA, Lynch CJ, Breeding RJ (1975) Vulnerability model. A simulation system for assessing damage resulting from marine spills. Springfield, Virginia: National Technical Infomntion Service, U. S. Department of Commerce
17. Tsao C K, Perry WW (1979) Modifications to the vulnerability model: a simulation system for assessing damage resulting from marine spills. Final report. United States: National Technical information Service, U.S. Department of Transportation
18. Wieczorek CJ, Dembsey NA (2016) Effects of thermal radiation on people: predicting 1st and 2nd degree skin burns. In: Hurley MJ (ed) SFPE handbook of fire protection engineering. Springer, New York, pp 2705–2737

Chapter 6
Application of Theoretical Framework into the Jet Fire Consequence of Nature Gas Transmission Pipeline

Contents

6.1 Pipeline Incident in Tangshan, China 112
    6.1.1 Gas Leakage and Jet Fire 112
    6.1.2 Radiant Heat Flux and Injury/Damage 114
6.2 Pipeline Incident in Ghislenghien, Belgium 116
    6.2.1 Gas Leakage and Jet Fire 117
    6.2.2 Radiant Heat Flux and Injury/Damage 117
6.3 Summary 123
References 124

The increasing demand of natural gas (NG) and the distances between sources and consumers necessitate the transportation of NG through pipelines. Pipelines, especially those underground, are the safest, most reliable, economical and eco-friendly mode of product transportation. Pipeline incidents have a low frequency of failure, but they can have high consequences in terms of cost. In addition, analysis of accidental data suggests that the probability of self-ignition caused by a leaked flow is proportional to the operating pressure (P) of the NG transmission pipeline and the square of pipeline diameter (D), as shown in Fig. 6.1. Therefore, it is crucial to predict the consequences of a jet fire accident following the failure or rupture of the NG transmission pipeline.

The immediate ignition of an NG pipeline leakage produces an explosive fireball that typically lasts less than 20 s, followed by the formation of jet fires. The duration of a jet fire is reported to reach up to 20 min [2]. Missiles and overpressure associated with the fireball can also occur simultaneously, but their hazard ranges are smaller than thermal radiation hazards posed by jet fires [3]. Full-scale experimental measurements and model calculations for the thermal radiation hazard from fireballs are widely available in the literature [3–5]. Consequently, this chapter focuses exclusively on applying theoretical framework to the thermal hazard assessment of the transient jet fire that occurs after the fireball has dissipated.

© The Author(s), under exclusive license to Springer Nature Singapore Pte Ltd. 2024   111
K. Zhou, *Jet Fire Due to Gas Leakage*, Springer Series in Reliability Engineering,
https://doi.org/10.1007/978-981-97-5329-1_6

Fig. 6.1 Ignition probability versus PD^2 for ruptured NG transmission pipelines (accidental data from Acton and Baldwin [1])

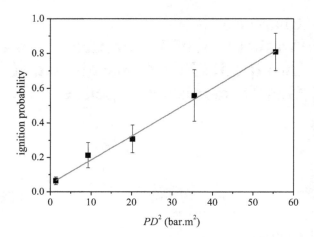

6.1 Pipeline Incident in Tangshan, China

On May 4, 2010, a fire and explosion occurred in the valve room of the pipeline segment from Tangshan to Qinhuangdao due to a lightning strike. Following the incident, the pipeline company cooperated with the local government to urgently evacuate nearby residents, shut off the two sectional valves in the upstream and downstream, release the natural gas from the pipeline, and suspend gas supply from the Qinhuangdao station to downstream areas. The vegetation on the adjacent hillside was also ignited, and the nearby residential buildings were destroyed. Fortunately, there were no casualties. The ruptured pipeline of 355.6 mm in diameter was connected to the main pipeline. The main pipeline had a diameter of 1016 mm and operated at a pressure of 5 MPa. The pipeline length was 24 km between the two remote-controlled sectional valves.

Figure 6.2 shows the zone heavily affected by the jet fire after the rupture of NG pipeline. The damage zone of the radiant heat reached as far as over 100 m. The pig was burned to death in the house 104 m away from the broken pipeline, and the grass was also burned to ash in the pig house. The window glass and metal frame were broken and melted, respectively, for the residential house 98 m away from the broken pipeline. All the indoor materials were also ignited for the residential house.

6.1.1 Gas Leakage and Jet Fire

The fully broken point was near one sectional valve. That is to say, the leakage exit was of 355.6 mm in diameter, and the total volume is that of the cylinder that holds the diameter of 1016 mm and the length of 24 km. the NG volume was 1994.77 m³. In calculation, the ambient air holds the temperature of 298 K in one standard

6.1 Pipeline Incident in Tangshan, China

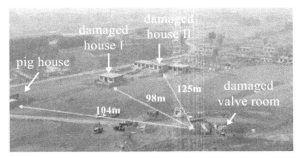

(a) zone heavily affected by the radiant heat from jet fire

(b) dead pig and burned small grass in the damaged pig house

(c) destroyed and indoor ignition of residential house I

Fig. 6.2 Radiant heat damage by the jet fire of pipeline incident in Tangshan

atmosphere pressure, and the temperature of NG is 293 K, and no wind is considered. These condition parameters are input into the theoretical framework for the calculation of the NG leakage rate, pressure, flame length and radiant heat flux field.

Figure 6.3 shows the prediction on the leakage flow velocity, the mass leakage rate and the density. The initial flow velocity and mass leakage rate can reach over 700 m/s and 1000 kg/s, respectively. As the leakage time goes, the leakage flow of NG undergoes from the choked and supersonic regime to the subsonic regime. Notice the use of the transient leakage model based on the van der Walls gas equation of state, for the leakage prediction.

Figure 6.4 presents the prediction on the flame length, the lift-off distance, the radiative fraction and the total heat release rate of the NG jet fire. The flame length reaches as long as 275 m in the initial time, while the lift-off distance is approximately 2.5 m. The maximum heat release rate approaches 50 GW. Notice the use of the flame length and radiative fraction correlations based on the flame Froude number, for the prediction of jet flame behavior.

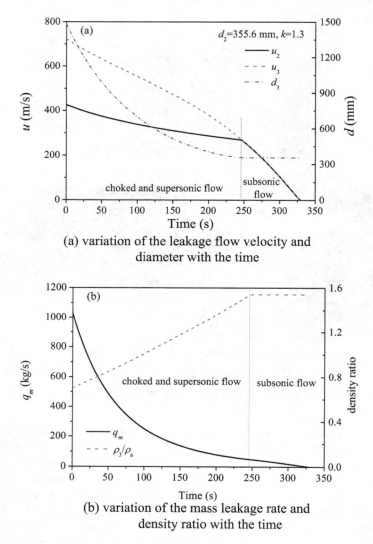

Fig. 6.3 Gas leakage prediction in Tangshan pipeline incident

6.1.2 Radiant Heat Flux and Injury/Damage

Figure 6.5 shows the prediction of radiant heat flux received by the target in the pig house. The target could be placed in any direction, e.g., the horizontal and vertical directions. The potentially maximum radiant heat flux is the square root of the sum of the squares of those in the horizontal and vertical directions. Notice the use of the line source radiation model, for the radiant heat flux prediction.

6.1 Pipeline Incident in Tangshan, China 115

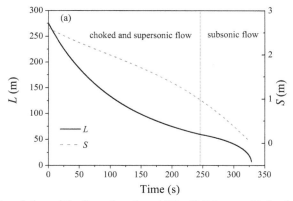

(a) variation of the flame length and lift-off distance with the time

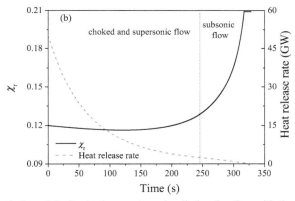

(b) variation of the heat release rate and radiative fraction with the time

Fig. 6.4 Jet flame prediction in Tangshan pipeline incident

Fig. 6.5 Prediction of radiant heat flux received by the target in the pig house

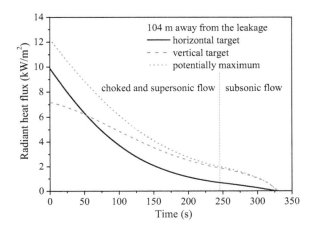

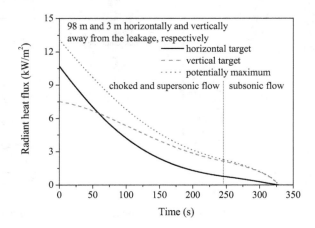

Fig. 6.6 Prediction of radiant heat flux received by the window of residential house I

Inputting the radiant heat flux curve into Eq. (5.4) can lead to the thermal dose. The radiant heat fluxes of the horizontal and vertical targets give the thermal doses of 1585 (kW/m^2)$^{4/3}$ and 1922 (kW/m^2)$^{4/3}$, respectively, while the maximum radiant heat flux results in that of 2828 (kW/m^2)$^{4/3}$, which is much larger than 1000 (kW/m^2)$^{4/3}$ s of the third degree burn threshold [14]. The comparison of thermal dose means the serious burn of the pig to death.

Inputting the thermal dose into Eq. (5.5) with $a = -12.8$ and $b = 2.56$ of Tsao and Perry [6], can lead to the probability of fatality. The radiant heat fluxes of the horizontal and vertical targets give the fatality probabilities of 85.54% and 94.06%, respectively, while the maximum radiant heat flux results in that of 99.46%. The calculation result indicates the high possibility of the pig's death.

Figure 6.6 shows the prediction of radiant heat flux received by the residential house I. The single glass would fail as exposed to the critical heat flux in the range of 4–5 kW/m^2 [7]. However, the glass thickness and installation form can affect the critical heat flux required for glass breakage. Wang et al. [8] argued that the thermal radiation from fires dominates the heat exchange mode to heat glazing in open space, and the crack of four different glass curtain walls occurs in the radiant heat flux range of 6–15 kW/m^2. As compared to the threshold of radiant heat flux to crack a glass, the predicted radiant heat flux in Fig. 6.6 can break the window glass of the residential house I.

6.2 Pipeline Incident in Ghislenghien, Belgium

On July 30, 2004, an accident occurred on a pipeline in an industrial area. The pipeline connected the port city of Zeebrugge with France. The technicians were able to isolate the pipe segment between the two sectional valves in two minutes after the explosion. A fireball occurred, accompanied by a large jet fire of 150–200 m in height. The jet

6.2 Pipeline Incident in Ghislenghien, Belgium

fire kept burning for approximately 20 min, until the gas supply was shut off. The accident caused 24 dead, 132 injured, and total devastation of an industrial zone over a 200-m radius, and the burning of many agricultural fields far away from the scene. The pipeline had a diameter of 1016 mm and operated at a pressure of 8 MPa. The pipeline length was 24 km between the two remote-controlled sectional valves.

In detail, an explosive crater formed of 14 m in length by 14 m in width by 4 m in depth. A packaging company located roughly 60 m away caught on fire. The wooden pallets were charred without ignition at an approximate distance of 130 m. The polyester cistern of rescue teams who stayed some 150 m from the scene, became warped due to the thermal radiation. The plastic recording box of a medium gas pressure reducing station, melted to cause a leak at a distance of 200 m. The green leaves of trees got scorching and the green grass burned at the distance of 210 m and 400 m away from the crater, respectively. Some literature [2, 5] reported the accidental pictures that apparently show the serious destroy and damage on people, building structure, cars, vegetations, etc.

6.2.1 Gas Leakage and Jet Fire

The fully broken point was between the two sectional valves. That is to say, the total volume was that of the cylinder that holds the diameter of 1016 mm and the length of 24 km. In calculation, the leakage exit is assumed to be of 200 mm in diameter, and the ambient air maintains a temperature of 298 K at one standard atmosphere of pressure, and the temperature of NG is 293 K, and no wind is considered. These condition parameters are input into the theoretical framework for the calculation of the NG pressure, mass leakage rate, flame length and radiant heat flux field.

Figure 6.7 shows the decrease of the leakage velocity and mass flow rate, as the initial pressure blows down or the leakage time goes. As shown, the leakage time is approximately 1200 s, which well agrees with the observation in the accident. The initial velocity of the leakage is over 700 m/s, and the mass flow rate approaches 600 kg/s. Notice the use of the transient leakage model based on the van der Walls gas equation of state, for the leakage prediction.

Figure 6.8 presents the decrease of the flame length and the lift-off distance with the leakage time. The initial flame length reaches 220 m, which also well follows the observation in the accidental process. The maximum heat release rate approaches 30 GW. Notice the use of the flame length and radiative fraction correlations based on the flame Froude number, for the prediction of jet flame behavior.

6.2.2 Radiant Heat Flux and Injury/Damage

Figure 6.9 shows the prediction of the radiant heat flux received by the packaging company located at 60 m away from the jet fire. The workshop building is assumed

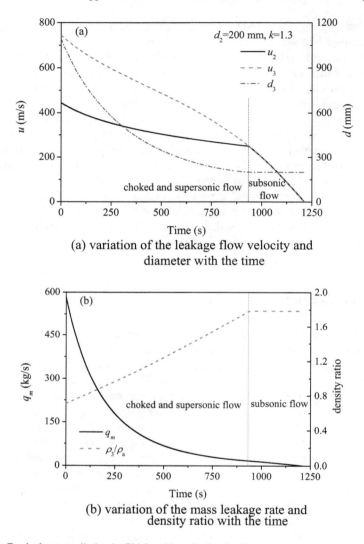

Fig. 6.7 Gas leakage prediction in Ghislenghien pipeline incident

to be 10 m in height. It should be stressed that the two sectional valves did not shut off until two minutes passed after the rapture. Thus, all physical parameters including the radiant heat flux, should keep the constant for two minutes before the decreasing as indicated in Fig. 6.9. The potentially maximum radiant heat flux reaches 23 kW/m^2 at the initial time, and it takes 40 s, 117 s, 202 s, 229 s and 287 s, to decrease to be 20 kW/m^2, 16 kW/m^2, 12.6 kW/m^2, 11.7 kW/m^2 and 10 kW/m^2, respectively. As compared to the thresholds of radiant heat flux in Table 5.3, most of polymer and wood materials could be ignited. In addition, Inputting the time profile of the

6.2 Pipeline Incident in Ghislenghien, Belgium

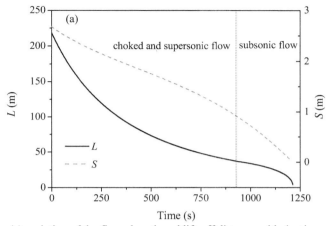

(a) variation of the flame length and lift-off distance with the time

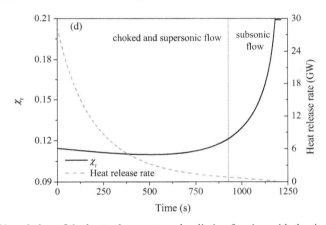

(b) variation of the heat release rate and radiative fraction with the time

Fig. 6.8 Jet flame prediction in Ghislenghien pipeline incident

potentially maximum radiant heat flux, Eq. (5.6) can give FTP = 11,731 kJ/m² larger than the FTP$_{ig}$, which indicates the ignition of wood by the radiant heat.

Figure 6.10 shows the time profile of the radiant heat flux received by the wooden pallet at the location of 130 m away from the leakage point. The maximum radiant heat flux is 7.8 kW/m² at the initial time, less than the threshold of 12.6 kW/m² required for wood ignition (see Table 5.3). Therefore, the wooden pallet cannot be ignited, even though the radiant heating effect of 6.4 kW/m² could last two minutes before the decreasing in the actual scene. However, the radiant heat can char the wood.

Figure 6.11 shows the decrease of the radiant heat flux with the leakage time, when the target, i.e., polyester cistern, is 150 m away from the jet fire. The

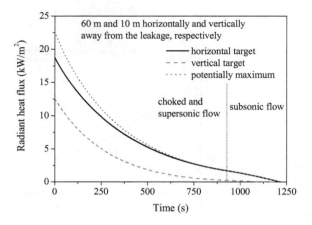

Fig. 6.9 Prediction of radiant heat flux received by the packaging company building

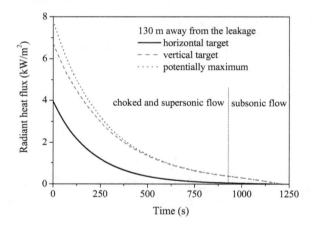

Fig. 6.10 Prediction of radiant heat flux received by the wooden pallet

maximum radiant heat flux is 6.4 kW/m² at the initial time, less than the threshold of 10 kW/m² required for polymer ignition (see Table 5.3). Therefore, the polyester cistern cannot be ignited, even though the radiant heating effect of 7.8 kW/m² could last two minutes before the two sectional valves were shut off. However, the radiant heat can increase the temperature of the polyester cistern and the water inside it. Accordingly, the polyester cistern only warped in the actual scene. The polyester cistern could melt if there was no water inside it.

Figure 6.12 shows the time profile of radiant heat flux received by the plastic recording box after the sectional valves shut off. The plastic box was 200 m away from the jet fire. The maximum radiant heat flux is 4.2 kW/m² at the initial time, less than the threshold of 10 kW/m² required for polymer ignition (see Table 5.3). Accordingly, the radiant heat cannot ignite the plastic box.

Table 6.1 lists the density, thermal conductivity and specific heat of typical plastic materials. The plastic recording box could be assumed to be the simi-infinite solid,

6.2 Pipeline Incident in Ghislenghien, Belgium

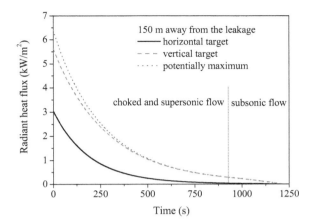

Fig. 6.11 Prediction of radiant heat flux received by the polyester cistern

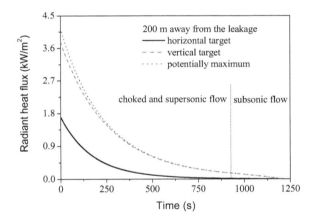

Fig. 6.12 Prediction of radiant heat flux received by the plastic recording box

thus Eq. (5.9) can calculate the temperature rise of different plastic materials under the radiant heating. Take the ambient temperature of 25 °C in summer for example. The final temperature can reach 188 °C, 222 °C, 239 °C, 296 °C and 333 °C, which is far more than the upper working temperature, for the CA, PC, PMMA, PP and PS, respectively. In calculation, the time profile of potentially maximum radiant heat flux is used, and the mean values of the physical properties are applied for the plastic materials. In addition, the melting temperature is 200–280 °C and 170–280 °C, respectively. Accordingly, the plastic box would fail, and even melt.

Figure 6.13 shows the time profile of radiant heat flux received by the green leaves of trees after the sectional valves were shut off. The trees along the road were 210 m away from the jet fire, and held 3 m in height. The maximum radiant heat flux is 3.9 kW/m^2 at the initial time. The minimum radiant heat flux is 500–800 kW/m^2 to ignite the combustible forest materials, i.e., the needles of cedar, pine, and fir-tree, birch leaves, lichen and moss [9], while it is 20–50 kW/m^2 for the ignition of dry

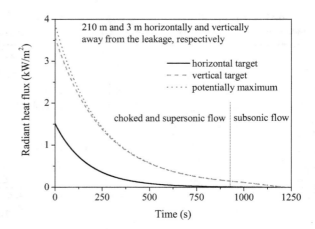

Fig. 6.13 Prediction of radiant heat flux received by the green tree leaves

plant leaves [10]. Accordingly, the radiant heat cannot ignite the tree crown in the jet fire accident.

Table 6.2 lists the thickness of the leaf sample, density and specific heat capacity of seven species of plant leaves. The variation of tree leaf temperature could be calculated by the lumped capacitance method. Thus, Eq. (5.10) can calculate the temperature rise of different plant leaves under the radiant heating. The ambient air temperature is also assumed to be 25 °C in summer for example. The temperature after radiant heating can reach 93 °C, 66 °C, 89 °C, 72 °C, 77 °C, 118 °C and 101 °C, for the Artocarpus, Jackfruit, Cinnamomum, Mangifera, Cocos, Tectona and Myristica, respectively. It should be notice that the calculated temperature could not be reasonable, for the fresh leaf holds high moisture content and the water evaporates to cool the leaf. Anyway, 60 °C is the lethal temperature of the live foliage to scorch tree crown [11]. Accordingly, the green leaves of trees only got scorching in the jet fire accident.

Figure 6.14 shows the time profile of radiant heat flux received by the green grasses that were 400 m away from the jet fire. The maximum radiant heat flux is 1.3 kW/m² at the initial time. The critical heat flux is 15–24 kW/m² for the ignition of three little

Table 6.1 Thermophysical properties of typical plastics

Material	Formula	Density kg/m³	Specific heat capacity J/kg·K	Thermal conductivity W/m·K	Upper working temperature °C
Cellulose acetate	CA	1220–1340	1200–1900	0.16–0.36	55–95
Polycarbonate	PC	1350–1520	1200	0.19–0.22	115–130
Polymethylmethacrylate	PMMA	1100–1200	1400–1500	0.17–0.19	50–90
Polypropylene	PP	900–910	1700–1900	0.1–0.22	90–120
Polystyrene	PS	1040–1050	1200	0.1–0.13	50–95

6.3 Summary

Table 6.2 Thermal properties of seven species of plant leaves [12] and the temperature after radiant heating

Plant species	Thickness of the leaf sample mm	Density kg/m³	Specific heat capacity J/kg·K	Temperature after heating °C
Artocarpus	0.28	631	1637	93
Jackfruit	0.33	650	2252	66
Cinnamomum	0.20	687	2267	89
Mangifera	0.21	879	2263	72
Cocos	0.32	918	1287	77
Tectona	0.20	475	2232	118
Myristica	0.24	866	1255	101

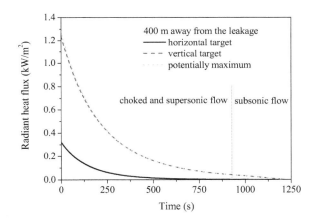

Fig. 6.14 Prediction of radiant heat flux received by the grasses of 400 m away from jet fire

bluestem grasses [13]. Accordingly, the jet fire is impossible to ignite the grasses at the position of 400 m away from the crater. The grasses near the crater should be ignited to induce the grass fire spread, but the fire spread stopped at the position of 400 m. The reason could be that the radiant heat cannot dry the grass far away. Take the thermal properties in Table 6.2 to simulate that of grass, the temperature rise is 1–2 °C, which indicates high moisture content still in the grass.

6.3 Summary

Two accidental cases involving natural gas transmission pipelines are quantitatively analyzed in terms of jet fire consequences. In detail, pipeline condition parameters are input into the theoretical framework depicted in Fig. 5.2. This framework is used to analyze the leakage process, jet flame, radiant heat, and the consequences of various targets. The calculated consequences of damage are well compared with those observed in the accidents, including the fatality, the cracking of window glass, the ignition of a packaging company building, the charring of wooden pallets, the warping of a polyester cistern, the melting of a plastic recording box, the scorching of green trees, and the ignition and spread of grass fires.

References

1. Acton MR, Baldwin PJ (2008) Ignition probability for high pressure gas transmission pipelines. In: Proceedings of the 7th international pipeline conference
2. ARIA. Rupture and ignition of a gas pipeline, July 30, 2004, Ghislenghien, Belgium: French Ministry of Sustainable Development - DGPR / SRT / BARPI No, 2009
3. Cleaver RP, Halford AR (2015) A model for the initial stages following the rupture of a natural gas transmission pipeline. Process Safety Environ Protect 2015: 202–214
4. Wang K, Liu Z, Qian X et al (2017) Long-term consequence and vulnerability assessment of thermal radiation hazard from LNG explosive fireball in open space based on full-scale experiment and PHAST. J Loss Prev Process Ind 46:13–22
5. Cowling N, Phylaktou H, Allason D et al (2019) Thermal radiation hazards from gas pipeline rupture fireballs. Proceedings of the proceedings of the ninth international seminar on fire and explosion hazards (ISFEH9), Saint-Petersburg, Russia, F, 2019. Saint-Petersburg Polytechnic University Press
6. Tsao CK, Perry WW (1979) Modifications to the vulnerability model: a simulation system for assessing damage resulting from marine spills. Final report. United States: National Technical information Service, U.S. Department of Transportation, 1979.
7. Mowrer FW (1998) Standards N I o, technology, Window breakage induced by exterior fires. National Institute of Standards and Technology
8. Wang Y, Wang Q, Su Y et al (2015) Fracture behavior of framing coated glass curtain walls under fire conditions. Fire Saf J 75:45–58
9. Grishin AM, Zima VP, Kuznetsov VT et al (2002) Ignition of combustible forest materials by a radiant energy flux. Combust Explos Shock Waves 38(1):24–29
10. Dahanayake KC, Chow CL (2018) Moisture content, ignitability, and fire risk of vegetation in vertical greenery systems. Fire Ecol 14(1):125–142
11. Zhou K, Simeoni A (2022) An analytical model for predicting the flame length of fire lines and tree crown scorching. Int J Wildland Fire 31(3):240–254
12. Jayalakshmy MS, Philip J (2010) Thermophysical properties of plant leaves and their influence on the environment temperature. Int J Thermophys 31(11):2295–2304
13. Overholt KJ, Cabrera J, Kurzawski A et al (2014) Characterization of fuel properties and fire spread rates for little bluestem grass. Fire Technol 50(1):9–38
14. Hockey SM, Rew PJ (1996) Review of human response to thermal radiation, HSE Contract Research Report No. 97/1996, Health and Safety Executive books, Sudbury, Suffolk, UK

Chapter 7
Conclusions and Future Research

Contents

7.1   Main Conclusions 125
7.2   Recommendations for Future Research 127
References 129

This book focuses on the dynamic process modelling of high-pressure gas leakage and jet fires. The aim is to quantify the thermal damage associated with the dynamic process. The widespread use of high-pressure gas and the associated risks of accidents are thoroughly discussed in Chap. 1. Chapter 2 introduces the transient leakage models for high-pressure gas. Chapter 3 discusses the correlations of flame length, lift-off distance and radiative fraction for jet fires, while Chap. 4 introduces the thermal radiation model for jet fires. In particular, Chaps. 2–4 compare the test results of 90 MPa hydrogen transient leakage and jet fires against the predictions of different models or correlations. Chapter 5 provides an in-depth review of the thermal damage criteria. Accordingly, a theoretical framework that integrates different models, correlations and criteria is proposed to accurately predict full-scale jet fires induced by high-pressure hydrogen or natural gas transient leakage. Finally, Chap. 6 applies the theoretical framework to analyze the consequences of jet fires in two natural gas transmission pipeline accidental cases.

7.1 Main Conclusions

(1) Conclusion on the transient leakage model

The transient leakage process of high-pressure gas can be predicted using the first law of thermodynamics, the equation of thermodynamic process, and the equation of state for gas. The predicted parameters encompass all state properties inside the leaked

© The Author(s), under exclusive license to Springer Nature Singapore Pte Ltd. 2024   125
K. Zhou, *Jet Fire Due to Gas Leakage*, Springer Series in Reliability Engineering,
https://doi.org/10.1007/978-981-97-5329-1_7

126   7   Conclusions and Future Research

vessel, the flow parameters and state properties at the leakage exit. The molecular volume and the intermolecular forces significantly affect the model accuracy, as illustrated by the comparison between predictions from ideal and real gas equations of state with experimental measurements. The high-pressure gas leakage evolves from an isentropic process to an isothermal process as the leakage time goes.

The vessel pressure could be sufficiently high to result in a choked leakage flow. The equation of local sound speed is needed to distinguish the transition from the choked flow to the subsonic flow. For the choked flow, a Mach disc appears downstream of the leakage exit, necessitating a notional nozzle model to correlate the gas flow parameters at the leakage exit with those at the Mach disc.

(2) Conclusion on the jet diffusion flame behavior

The available correlations of flame length, lift-off distance and radiative fraction can be coupled into the transient leakage model. The predictions of transient flame behavior against the measurement show the superiority of the flame length and radiative fraction correlations based on the flame Froude number. In addition, the summary reviews are also conducted for the effect of the cross wind, leakage exit shape, obstacle, solid particle, pit and underwater on the jet flame behavior. The predicted transient flame length, lift-off distance and radiative fraction can be input into a thermal radiation model.

(3) Conclusion on the jet flame radiation model

The point source radiation model, multipoint source radiation model, solid flame radiation model and line source radiation model are introduced in detail. The predictions of radiant heat flux against the measurement are conducted for the small and medium vertical jet fires, medium horizontal jet fires and small inclined jet fires. The comparisons between the four models show the superiority of the line source radiation model. The line source radiation model can be coupled into the jet flame behavior correlations and the transient leakage model, for the good prediction of radiant heat flux field of jet fires during the 90 MPa hydrogen transient leakage.

(4) Conclusion on the theoretical framework for calculating jet fire risk induced by high-pressure transient leakage

The theoretical framework consists of the gas leakage model, jet flame behavior correlation, thermal radiant model and thermal damage criteria. The thresholds of radiant heat flux, thermal dose and target temperature can be used as the thermal damage criteria. The methods are introduced to calculate the thermal dose and target temperature with the known transient radiant heat flux, for people injury and structure destroy.

The theoretical framework is justified by three field test results of the hydrogen, natural gas, and hydrogen/natural gas mixture jet fires due to pipeline leakage. The input parameters by importance are the leakage exit area, the pipeline pressure, the ambient pressure, the pipeline length and the ambient wind speed for the theoretical framework. The calculation uncertainty of the theoretical framework can be quantified by the lognormal density function. The theoretical framework can accurately calculate the consequences of damage in actual leakage incidents of nature gas transmission pipeline.

7.2 Recommendations for Future Research

High-pressure gas leakage often results in jet fires that cause secondary accidents. It can be typically divided into three consecutive processes: the high-pressure gas leakage flow, the jet flame combustion, and the thermal damage due to the heat from the jet fire. The academic community has conducted in-depth research on the leakage dynamics of high-pressure gas, the combustion behavior of jet fire, the heat transferred from the jet fire to the surroundings, and the thermal damage criteria of different objects. However, most of existing studies independently consider the three consecutive processes, with the development of corresponding models. The models can be concatenated to quantify the consequence of jet fires after high-pressure gas leakage, but they could cause errors gradually amplifying during the transmission between different models. Accordingly, it is necessary to select appropriate models and consider the coupling matching between different models, by analysis of specific leakage and combustion conditions. The prediction uncertainty challenges the available models of high-pressure gas leakage and jet fires.

The theoretical foundations for modeling high-pressure gas leakage are the first law of thermodynamics, the thermodynamic process equation, and gas equation of state. However, the model development requires three assumptions that could lead to the prediction uncertainty. The first assumption is an isentropic process for the leakage process. In fact, leakage is a non-adiabatic process involving heat loss or gain. For example, high-speed friction at the leakage exit results in heat dissipation, and heat exchange between the surrounding air and the gas inside the vessel could occur through the vessel wall. For high-pressure and high-temperature gas leakages, the transient decay of pressure inside the tank predicted by the isentropic process equation, can be significantly corrected by considering heat loss [1]. The second assumption is the constant specific heat capacity or specific heat ratio, independent of the gas temperature. In fact, the specific heat ratio varies with gas temperature and pressure. Accordingly, this assumption can lead to deviations and even failures of model calculation, when there is significant temperature variation during the leakage. The third assumption relates to the equation of state that describes the gas behavior. Air behaves as a non-ideal gas, when the air pressure exceeds 50 standard atmospheric pressure [2], for the important effects of molecular volume and intermolecular forces. The Abel-Noble equation of state only considers molecular volume, while the van der Waals equation of state takes into account both molecular volume and intermolecular forces. However, existing real gas equations of state are derived by introducing correction factors into the ideal gas equation of state. The correction factors are empirical to inevitably result in the certain adaptation condition or limitation for the equation of state. The three assumptions cause the uncertainty for

128   7 Conclusions and Future Research

high-pressure gas leakage models, which requires an estimation of deviation trends and levels by analysis of actual factors affecting the models.

The model of notional nozzle is derived from the conservation equations of mass and momentum with several assumptions. The first assumption is no air entrained into the jet flow through the boundary of the notional nozzle, while the second one is the neglect of the viscous forces at the boundary of the notional nozzle. In the third assumption, the pressure at the Mach disk equals the ambient pressure, and the temperature at the Mach disk equals either the ambient temperature or the temperature inside the tank. The first two assumptions help to establish conservation equations of mass and momentum between the leakage exit and the Mach disk, while the third one helps to close the conservation equations. The velocity profile at the leakage exit is parabolic, whereas it is uniformly distributed at the Mach disk. This necessitates the introduction of a leakage coefficient to enhance the reliability of model prediction. Some scholars suggest a sonic velocity model to calculate the effective velocity at the Mach disk and then the mass conservation equation to calculate the effective diameter at the Mach disk [3], which abandons the momentum conservation equation. In essence, different calculation methods for the physical parameters at the Mach disk imply different definitions of the notional nozzle. The difference poses a challenge of uncertainty in quantitatively describing supersonic leakage.

The modeling or model validation of flame geometry mainly relies on experimental data of subsonic jet fires, which introduces uncertainties of these models used for supersonic jet fires. Many scholars have studied the attenuation of velocity and concentration along the axis of subsonic jet flow and established the attenuation models. The attenuation models are used to well predict the concentration [3] and velocity [4] of supersonic jet flows, if the flow parameters at the Mach disk are the input parameters of model. Similarly, Schefer et al. [5] directly input the effective diameter and flow velocity at the Mach disk into the subsonic flame geometric model, for the prediction of supersonic jet fire behavior against experimental measurement. Clearly, merely replacing input parameters in subsonic models does not alter the principle that the flame geometry is primarily influenced by the Froude number. However, compared to subsonic jet fires, Reynolds number and Mach number also remarkably affect the length of supersonic flames [6]. Therefore, the future research needs to further investigate the coupled effects of Froude number, Reynolds number, and Mach number on supersonic jet fires.

The factors considered in the radiative fraction correlation are not comprehensive. Equations (3.10) and (3.11) take into account the physical and chemical properties of gas fuel, and Eq. (3.11) also considers the size of leakage exit and the flow velocity at the leakage exit. However, Eqs. (3.10) and (3.11) have significant limitations, for boundary conditions such as ambient wind speed, ambient pressure, and pit can also affect the radiative fraction of jet fires. The maximum radiative fraction is 0.1 for pure hydrogen jet fires [7], while it can reach 0.29 when the sand is entrained into the hydrogen jet fires [8]. But the sand significantly reduces the radiation fraction of propane jet fires [9]. The pit caused by the explosion of underground pipelines, increases the radiative fraction of propane jet fires [10]. For Case 2 in Sect. 5.3.2, Eq. (3.11) may coincidentally predict accurately the radiation fraction of natural gas jet fires, possibly because the negative effect of sand counterbalances the positive effect of pit. Obviously, there are numerous potential factors influencing jet fire radiative fraction, which requires objective and discriminative analysis to estimate the deviation trends and levels of radiative fraction caused by the factors. Furthermore, experimental measurement is scarce on the radiation fraction of supersonic jet fires.

In addition to the three consecutive processes of high-pressure gas leakage flow, jet combustion, and thermal damage caused by the heat from the burning flame, it also needs to analyze the thermal feedback from the jet fire to the leaked vessel. The analysis aims to understand the behavior of the pressure and temperature inside the vessel resulting from the heating effect of thermal feedback. It ultimately achieves simulation of the entire sequence of evolving events: "high-pressure leakage flow $\rightarrow$ jet flame combustion $\rightarrow$ thermal feedback $\rightarrow$ tank temperature/pressure rise $\rightarrow$ intensified leakage flow".

References

1. Enkenhus KR (1967) On the pressure decay rate in the longshot reservoir. Von Karman Institute for Fluid Dydamincs
2. Donaldson CD (1948) Note on the importance of imperfect-gas effects and variation of heat capacities on the isentropic flow of gases. Washington, DC: NACA 1948, RM No. L8J14: 1–21
3. Birch AD, Brown DR, Dodson MG et al (1984) The structure and concentration decay of high pressure jets of natural gas. Combust Sci Technol 36(5–6):249–261
4. Birch AD, Hughes DJ, Swaffield F (1987) Velocity decay of high pressure jets. Combust Sci Technol 52(1–3):161–171
5. Schefer RW, Houf WG, Williams TC et al (2007) Characterization of high-pressure, underexpanded hydrogen-jet flames. Int J Hydrogen Energy 32(12):2081–2093
6. Molkov V, Saffers J-B (2013) Hydrogen jet flames. Int J Hydrogen Energy 38(19):8141–8158
7. Zhou K, Qin X, Wang Z et al (2018) Generalization of the radiative fraction correlation for hydrogen and hydrocarbon jet fires in subsonic and chocked flow regimes. Int J Hydrogen Energy 43(20):9870–9876
8. Acton MR, Allason D, Creitz LW et al (2010) Large scale experiments to study hydrogen pipeline fires. In: Proceedings of the 8th international pipeline conference, Calgary, Alberta, Canada
9. Zhou K, Nie X, Wang C et al (2021) Jet fires involving releases of gas and solid particle. Process Saf Environ Prot 156:196–208
10. Zhou K, Zhou M, Huang M et al (2022) An experimental study of jet fires in pits. Process Saf Environ Prot 163:131–143

Printed in the USA
CPSIA information can be obtained
at www.ICGtesting.com
CBHW070340230924
14770CB00004B/222